# ASTROQUIZZICAL

Unipress Books

www.unipressbooks.com
First published 2022
Design and Layout © UniPress Books Ltd 2022
Text copyright © 2022 Jillian Scudder

Published by MIT Press by arrangement with UniPress Books Ltd.

Commissioning Editor: Kate Shanahan
Project Manager: Richard Webb
Design & Art Director: Paul Palmer-Edwards
Illustrator: Robert Brandt
Picture Researcher: Richard Webb

ISBN 978-0-262-04672-5

Library of Congress Control Number 2021937296

Printed in China

The MIT Press
Massachusetts Institute of Technology
Cambridge, Massachusetts 02142
http://mitpress.mit.edu

# ASTROQUIZZICAL

## SOLVING THE COSMIC PUZZLES
## OF OUR PLANETS, STARS,
## AND GALAXIES

### THE ILLUSTRATED EDITION

**JILLIAN SCUDDER**

The MIT Press
Cambridge, Massachusetts

# CONTENTS

# INTRODUCTION

*astroquizzical*
**adj.** *expressing curiosity in the astrophysical wonders of the universe*

It is often said by astrophysicists that every one of us should feel a strong connection to the stars. Without generations of stars that burned, exploded, or collided before our planet was formed, the carbon that our bodies are made of, the iron in our blood, and the gold and silver of our precious objects would simply not exist. In a very tangible way, those stars made it possible for us to be here to look at them. Without them, we could not possibly have evolved on our watery world. But truly exploring how we are linked to them—and how they have led to our own lives on planet Earth—can be an arduous task, even for the curious-minded among us. While there are many ties between us and the stars, such information is often forgotten or hard to find.

This book explores the ties that link us not just to the stars, but to the universe as a whole—our cosmic family. Without our parent planet to call home, we would not exist. Without a star, the Sun, our planet would not exist. Without a galaxy, our star would not exist. And without the filamentary nature of structure in the earliest universe, our galaxy would not exist. Each of them paved the way for another generation, building up the groundwork for our tree of life.

Welcome to our cosmic family tree! We are living, in company with almost every other human, on the surface of the planet Earth, the only planet in the vast universe known to host intelligent life of any kind. And we are doing so because billions of years ago the Big Bang created the atoms that, over unimaginable periods of time, formed the stars and galaxies that we can still see from our home planet. In this book we will use this family tree structure to travel out into space and at the same time back in time through the cosmic generations, from our Earth (parent) and its sibling planets to the Sun and stars (grandparents), the Milky Way and other galaxies (great grandparents), and on to the very creation of our universe (great-great grandparents). This will help us to discover our unique place in the cosmos, and to understand a little more of what it really means to stare up into the night sky and gaze at the stars.

This journey will also encourage us to be more "astroquizzical," by asking questions of our universe and its ancient stars, conducting "thought experiments" that explore the workings of our planet and its distant relatives, and seeking answers to the many mysteries of space from the clues we can discover at the very beginning of time.

# TIMELINE OF THE UNIVERSE

**0 years**  The Big Bang: the universe comes into existence

**10 seconds–15 minutes**  The first atoms form

**375,000 years**  The Cosmic Microwave Background is produced (the universe becomes transparent to light)

**c. 180 million years**  The first stars in the universe form

**c. 200 million years**  The oldest stars in the Milky Way form

**400 million years**  The oldest observed galaxy (to date)

**3.5 billion years**
**(10.3 billion years ago)**  Stars form in galaxies at the highest rate in cosmic history

**9.2 billion years**
**(4.57 billion years ago)**  The Sun forms

**9.3 billion years**
**(4.47 billion years ago)**  Earth forms

**13.77 billion years**  "Now," relative to the start of the universe

# THE SKY
# FROM HOME

We humans live atop a delicately positioned planet, and throughout our history we have watched the stars. While we now aspire to land humans on other planets, we have learned a considerable amount about the universe simply from observing our own skies.

# WHAT CAN WE REALLY SEE OF THE UNIVERSE FROM EARTH?

When hunting for our own human ancestors, it makes sense to start at the top of the family tree—with us. Similarly, we begin our journey up the cosmic family tree with the planet we call home, Earth.

How do we see the universe from where we stand? As both the children of this vast cosmos, and the only ones (to our knowledge) who are attempting to chart our cosmic lineage and understand the rules of the universe, our own perspective is a unique one. Our sense of what the "big picture" is, as well as how we might fit into that picture, is very much affected by how the stars appear to us. As we have developed more advanced ways of viewing the night sky, our sense of just how big the picture truly is has only grown in scope.

Our view of the cosmos is almost always from the surface of our parent planet, Earth. A select few members of our human race have had the privilege of observing our planet from a loftier perch, but with the exception of those astronauts, all of humanity has observed the stars and planets from the ground. It's at night, with our atmosphere protecting us from the freezing void of space, and our own star no longer flooding our planet with light, that we can see the first glimpses of our immediate cosmic family.

On a clear night, many of us will look up to find ourselves briefly captivated by the shining of a bright object in the sky. Some of the brighter lights we can see in the night sky are Earth's planetary siblings, formed out of the exact same cloud of dust and gas that generated our own home 4.5 billion years ago (there is more on this in Chapter 3). But even without the planets overhead to shine extra brightly, the night sky can be dazzling, especially if you happen to find yourself away from the lights of the city streets.

## ACCOUNTING FOR EARTH'S ROTATION

To capture the faintest stars in an image where many thousands of other stars will appear, you have to take into account the rotation of Earth. Earth rotates every 24 hours, of course, and if you want to take images of a single set of stars over the course of several hours, they will be moving dramatically as Earth rotates us. To counteract this, many of the deepest images are taken by attaching the camera to a mount that can pivot as Earth turns, constantly correcting for the spin of the planet. With this technique, you can take even more images to assemble together, allowing you to bring out the light from fainter and fainter stars as you spend more time taking photos.

But just as the images from the International Space Station can remind us of the curvature of Earth, photographers who have the means to travel to the few remaining truly dark places, where there is little or no light pollution (see the box, opposite), can capture the night sky in those remote spots, reminding the rest of us of what we're missing.

▲ A meteor streaks across a star-filled night sky in Coconino National Forest near Flagstaff, AZ.

## LIVING WITH LIGHT POLLUTION

Without interference from artificial lights, thousands of stars are visible even to our unaided eye, but these skies are increasingly difficult to find. According to a 2016 study by Fabio Falchi, 99 per cent of the US and European populations live under light-polluted skies. It's easy to forget, or to have never seen, just how many stars are visible to us from Earth.

## CAPTURING THE STARS

Many photographers aim their lenses at the Orion Nebula. It's both a very esthetically pleasing part of the sky, and a very bright one, so it's easy to capture a number of stars in the image. Looking at some of these photographs of the night sky, and then looking at the version above your own homes, it may seem that the images have been exaggerated somehow, or that the number of stars has been digitally increased. This isn't the case—a camera has an advantage that our own eyes can't access. Our eyes are relatively small light-capturing devices, and we can't increase the exposure time in the same way that you can on a camera to catch even faint light.

Many photographs of the night sky (such as the image on page 11) are not exaggerated, but come from a very long time spent observing the sky with a much larger lens than our eye. The longer you point your camera at a specific part of the night sky, the fainter the starlight you're able to capture. Once the light has been collected, we arrive at an astounding view of what our night sky looks like, beyond the limitations of light pollution and the small size of the human eye.

No special filters are required here; many of these photographs are taken by regular digital cameras—slightly fancier than the one in your phone. The astronauts on the International Space Station may have a unique position from where they can take photographs, but they use the same technique as the photographers on the ground. If you take a series of 10- to 30-second photos of the sky, you can assemble your series of images into a single, much more detailed record. The stars that exist in all of the exposures should pop up more brightly in the assembled image, and anything that happens to show up in one 30-second window but not another, will fade.

The longer you point your camera at a specific part of the night sky, the fainter the starlight you're able to capture.

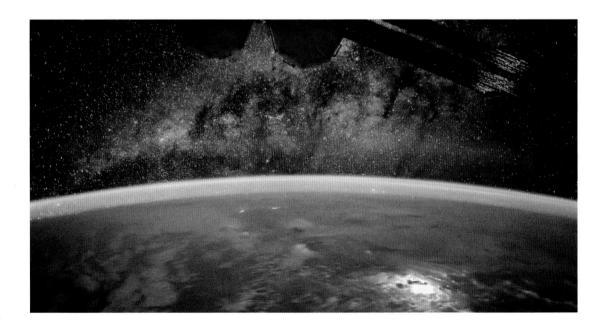

◄ Taken by an astronaut aboard the International Space Station (ISS) in 2015, this oblique view of Earth features a thunderstorm. The glare from lightning reflects off the solar panels of the ISS. In the background are the stars and dust clouds of the Milky Way. The red curved band passing through the center of the image is the faint glow of an upper layer of our Earth's atmosphere.

▲ The Orion Nebula, as captured by the Hubble Space Telescope, after 168 hours of observation. The Orion Nebula is a picture book of star formation, from the massive, young stars that are shaping the nebula to the pillars of dense gas that may be the homes of budding stars. The bright central region is the home of the four heftiest stars in the nebula, which are known collectively as the Trapezium because they are arranged in a trapezoid pattern. Ultraviolet light unleashed by these stars is carving a cavity in the nebula and disrupting the growth of hundreds of smaller stars. This image represents approximately the same area of the sky as is occupied by the full moon.

THE SKY FROM HOME

13

# WHAT COLOR IS THE UNIVERSE?

Images of our cosmic relatives come to us from beyond our planet's surface. Pictures from the Hubble Space Telescope, for instance, have revealed that the world beyond Earth is a vivid and highly detailed one. But this vividness can present a puzzle.

The human eye has an unusual sensitivity pattern to light. We're pretty good at seeing things in the yellow-green range, but once you get into reds and blues our eye suddenly gets extremely bad at registering deep reds and dark purples, and our brain translates those colors into "black" or, more accurately, as "there is no light here that I can deal with." To anything outside the range of visible light, we are completely blind. This odd sensitivity pattern means that it's quite difficult to make a camera with exactly the same sensitivity as our eye.

If you want to make a color picture from an image coming from a space telescope, there is a further challenge to overcome. All the cameras attached to telescopes are just photon counters (the photon is the smallest quantity of light—one individual packet or particle). If a photon makes it through the telescope and into the camera, it adds one to the number of photons that arrive from that patch of the sky. This means that the only images you can make are intensity maps—black-and-white images.

For scientific purposes, astronomers are generally more interested in measuring the amount of very specific color slices of light that arrive at the telescope. In order to limit the kind of light that actually makes it to the telescope's camera, filters are usually put in front of it. The filter works in the same way as red-blue 3D glasses and images: the red lens lets through only the red light, and the blue lens lets through only the blue, so each eye gets a different picture, and your brain reconstructs the depth of the image.

An astronomical filter is usually constructed to let in light from a very specific physical process— for instance, the color of light that hydrogen produces when it is near a very hot star. Hydrogen here creates a deep-pink color, so instead of a red or blue filter we'd have a deep-pink one. This would let in only light that is produced by hydrogen, and we can map the locations of that gas on the sky. This image is still entirely in black and white, but it allows astronomers to understand what's happening in that part of the sky.

## ARE "FALSE-COLOR" IMAGES FALSE?

To reconstruct a colorful image out of this black and white one is no simple task. Given that we're detecting light at much greater sensitivities than the human eye, and that we're usually doing it in discrete chunks instead of one (very complex) curve, as the eye does, putting these chunks of light back into a single image is a tricky business. Even when all the light is taken from the narrow range that we can see, it must still be reconstructed and tweaked to reflect the brilliance of the colors we've observed. While the general term for this style of image is "false color," the colors here aren't actually "false." The deep-pink glow of hydrogen will remain deep pink, and the glow of oxygen, a brilliant aqua, will stay that color.

## BEYOND OUR SIGHT

"Exaggerated" color images can be used to extend our sight far beyond what we can actually see. Perhaps a galaxy is rather unimpressive in visible light, but has a stunning brilliance in the ultraviolet or X-ray. To our eyes, this is dark; but a black-and-white image from a telescope sensitive to that light can be added to our collection, allowing us to construct an image. In these cases, a color too blue for human eyes is often added as a vivid blue or purple, and a color too red for us is added as a bright red or purple.

▲ Hubble captures billowing clouds of cold interstellar gas and dust rising from a tempestuous stellar nursery located in the Carina Nebula, 7,500 light-years (see the box on page 91) away. The image is rendered with six colors, each color coming from its own black-and-white photograph of the sky. In other words, the images come out of the telescope as black and white, but each is assigned a color and then reassembled. The colors in the composite image correspond to the glow of oxygen (blue), hydrogen (pink), nitrogen (green), and sulfur (red).

# WHY DO STARS TWINKLE?

It's easy to forget that from the ground, our view of our cosmic family is strongly influenced by the presence of our atmosphere. Even when the skies are clear, our atmosphere can pose some barriers to seeing the stars as clearly as we might like.

The atmosphere on Earth is transparent to the light most of us use to navigate through our world, but that doesn't mean that light remains unchanged as it encounters our atmosphere. If you've ever seen the stars twinkling overhead in the view from home, you've witnessed one of these transformations.

When the stars overhead appear to flicker and change in brightness, that's the atmosphere at work, and the technical term to describe this flickering is scintillation. The stars themselves are quite stable, and the amount of light they produce and send toward our little planet isn't changing. We can check this by waiting a few days—if you go out on a clear, still night, you should find that the stars hang quietly in the night sky, not a twinkle to be spotted. And if you were fortunate enough to observe the stars from space, you would see them as perfectly pointlike pricks of light, no matter how twinkly they appeared from the ground. Back on Earth, if the wind has picked up, you should be able to see the stars twinkling their hearts out. You might see something similar if you look to the stars near a low horizon (sorry, city dwellers). Even if the stars directly overhead seem stable, the ones closer to the horizon may appear unsteady.

Whenever light encounters a gas—and our atmosphere is all gas—the direction in which the light points changes slightly. How much the light bends depends on a number of things, but the density of the gas is one of them. How densely packed the gas is depends strongly on its temperature, so warm air bends light to a slightly different angle than cool air does. The higher you are in the atmosphere above our planet, the cooler the temperature, but this isn't a completely smooth transition from warm to cold. These temperature changes exist in little bubbles of air, packed against one another. These little bubbles act like a series of lenses suspended above us, twisting and distorting the light on its way through. If the air is calm, these air-pocket lenses are relatively large, so the light travels through fewer of them, and has fewer deflections on its way down to the ground. Similarly, if the temperature isn't changing rapidly from bubble to bubble, the light won't change direction as often.

For this reason, stars overhead might appear to twinkle less than stars at the horizon. Light from a star directly overhead is taking the shortest path through our atmosphere: straight down. A star close to the horizon is taking one of the longest paths possible, and so the number of air lenses it must travel through in order to reach our eyeballs is much higher (see the diagram opposite). With more chances for the starlight to be bent into an unusual place, the likelihood that the starlight will flicker in and out of focus goes dramatically up— these atmospheric focusing problems are what we see as a twinkle in the night skies.

## ARE WE SEEING THINGS STRAIGHT?

The technical term for this atmospheric interference with starlight is "seeing." The better the seeing, the less twinkling the stars will do. Even on a perfectly clear night it's possible to have bad seeing, usually due to wind in the atmosphere. If you're after extremely crisp images of objects far away from our parent planet, bad seeing can be a significant problem—with the atmosphere warping pinpricks of light into much larger, indistinct shapes, the images that come out of the telescope are more blurry than we would like. If you're trying to distinguish two closely placed stars, bad seeing can blur them together. This is one of the main reasons why astronomers like space telescopes: there's no blurring of the distant starlight in space.

## ► Atmospheric distortions at work

Stars that are overhead travel through less of Earth's atmosphere and are therefore less distorted by pockets of hotter and cooler air. Stars that are near the horizon are subject to more atmospheric distortion as their path through the air is much longer. In addition to poor seeing, these horizon stars also have their apparent position altered, as the light from the stars is bent as it travels through the atmosphere. This has the effect of making them appear slightly higher in the sky than they actually are (this effect is exaggerated in the diagram below).

## Scintillation

Pockets of warmer (less dense) and cooler (more dense) air are in motion, constantly altering the path of light from stars very slightly.

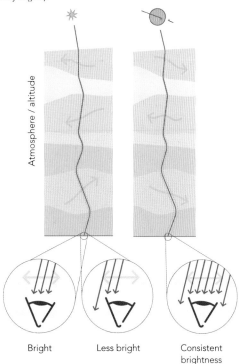

Atmosphere / altitude

Bright · Less bright · Consistent brightness

The amount of light hitting the observer's eye varies, creating the impression of a fluctuation of brightness. Planets are physically larger in the sky, so small shifts in the way light travels through the atmosphere make less of a change to their brightness.

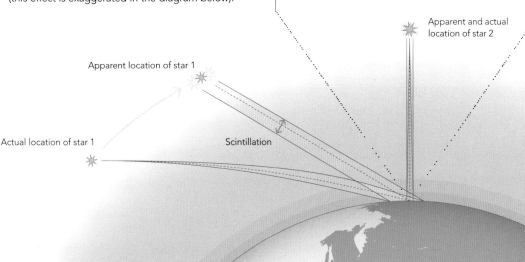

Apparent location of star 1

Actual location of star 1

Scintillation

Apparent and actual location of star 2

# WHAT ARE METEORS?

Meteors are little pieces of material, usually pebble-sized pieces of rock, that have the misfortune of running into our planet. The name meteor distinguishes them from objects in space, which are meteoroids, and bits of rock that actually survive the passage through the atmosphere and reach Earth's surface, which are called meteorites.

Running into the atmosphere of our planet spells doom for most small objects. The change from the void of space to the relatively high density of the gases of our atmosphere means that these pieces of the solar system are rapidly slowed down, like an arrow burrowing into a straw target. As they slow, they donate energy to the gas surrounding them, which heats up and vaporizes the outer layers of the rocks as they plunge groundward. If a meteor is small, this process will evaporate it entirely in the blink of an eye, and the flash of its glow fades from sight. This can happen at any time of day, but we associate it more with the night skies, because a meteor's luminous end is much easier for us to spot when the light from the Sun isn't there to obscure it. However, if the meteor is big enough, it won't matter whether the Sun is up. The fireball that exploded over Chelyabinsk, Russia, in 2013 was clearly visible in the morning sunlight.

There are particular times of year when the odds of spotting a meteor are much higher than normal—during what we call meteor showers. The odds go up because Earth is passing through a particularly pebble-filled patch of our orbit, the rocks usually being the remains of a comet that passed by many years earlier. As the comets, which are made of rock and ice, come close to the Sun, the ice melts, freeing the rock into space. These small rocks and other pieces of debris remain behind the comet's path, the way a particularly muddy dog is easily traced through a house. When Earth catches up with this path, we get a very small atmospheric pummeling, and a fun light show in the night sky.

▼ The smoke trail produced by the Chelyabinsk fireball's path through the atmosphere in 2013. The trail came from the high-temperature evaporation of the incoming rock.

Meteoroid

Meteor

Meteor
shower

The Canyon Diablo meteorite
crater in the United States.
Caused by an iron meteorite
50,000 years ago, it is 0.7 miles
(1.13 km) in diameter and
560 feet (171 meters) deep.

▶ **Meteoroids, meteors, and meteorites**
Before any rocky or metallic object
encounters Earth it's dubbed a meteoroid,
but it could become a meteor, if its timing
and aim are good. Meteors are the objects
we see as streaks of light in the atmosphere.
If the object that created the meteor
survives its journey through the atmosphere,
the remaining rock is called a meteorite.
Meteorites are always smaller than the
meteoroid they started as, since much of
their exterior will have been vaporized on
the way down to the ground.

# WHAT IS THE AURORA?

A meteor shower is certainly a dramatic nighttime event, but it is outdone by the aurora—the Northern or Southern Lights. Our planet doesn't have a monopoly on the aurora—they can be seen on other planets that are surrounded by magnetic fields. But even though we're one of a family of planets with the aurora, ours certainly puts on a delightful spectacle.

Only a few things are needed to produce the aurora: dark skies, a magnetic field, and an active star. Our home planet comes equipped with a magnetic field, generated by the motions of the metals in Earth's core, and dark skies come every 12 hours or so. An active star is also a regular occurrence, and what we need from our star, the Sun, is a large volume of charged particles (electrons or protons will do).

If the Sun provides these high-energy particles so that they hit Earth, our magnetic field will deflect the majority of them away from the planet before they ever come in contact with the atmosphere. However, if the charged particles travel toward Earth's poles, the magnetic field is less able to keep them away from the atmosphere. Near the magnetic poles, the magnetic lines that normally run parallel to the surface turn and sink into the surface (see page 107 for an illustration of this).

This creates a divot in the magnetic shield, and particles can get stuck in there, like leaves in an eddy. The solar particles end up smashing into the gas of our atmosphere. The energy donated to the atoms (see the box, opposite) of gas makes a glow, lighting up the skies with the aurora. The ring of light produced by the aurora describes an oval around the north or south poles, and is known as the auroral oval.

▶ The aurora is brilliantly visible in this image. Different colors represent different atoms in the atmosphere; this green is the particular signature of oxygen in our atmosphere.

▼ **Earth's magnetic field**
The Earth's magnetic field, drawn here in purple, is a literal force field surrounding our planet. The solar wind aimed at our planet is bent so that it travels away from our atmosphere instead of through it.

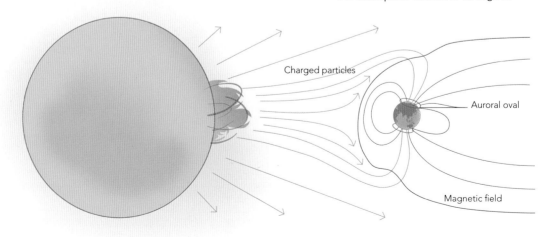

Charged particles

Auroral oval

Magnetic field

## HOW TO SPOT THE AURORA

If you're particularly keen to see the aurora, there are a few things you can keep an eye out for. As with all faint astronomical displays, it is easier to spot the aurora at night, and the darker your skies, the more visible it will be. The other factor is how close you are to the magnetic poles. The closer you are, the smaller the donation of particles needed from the Sun in order to see the lights. With very strong storms of particles, the aurora can be visible farther from the poles, but these storms are few and far between.

If it's starting to sound to you like all the good astronomical vistas from our parent planet are visible only when our illustrious Sun isn't around, that's not always the case. The Moon is easily visible in the daytime sky, and if you know where

to look, Venus is also bright enough to be seen while the Sun is up (various apps for your phone can point you in the right direction). There's nothing special about the Moon and Venus here, they're simply reflective enough to be visible—in principle, any kind of object in our skies could be seen from the ground during the day, assuming it can reflect enough light down to the surface.

### ATOMIC BUILDING BLOCKS

An atom is the smallest uncharged particle in the universe. Combinations of atoms make up all the materials on our planet and in the rest of the universe.

# If humanity built large spaceships in orbit around Earth, would we see them during the day?

Humanity hasn't yet made craft capable of transporting a large number of humans across interplanetary space, so let's go with something that does exist. At more than 984 feet (300 meters) long, the world's largest aircraft carriers (known as "supercarriers") are huge. If we replaced the steel that they are built from with the more reflective aluminum, would this be visible from the ground?

There are two things you need to consider when determining whether something is visible during the day:

1. How reflective is it? A dark, matte material will be much less reflective than something made of a shiny metal.

2. How big is its surface? The bigger the surface, the more light it will bounce from the Sun toward the surface of Earth, and the more easily visible it will be.

## 1. Reflectivity

We can use the International Space Station (ISS) as a benchmark. It reflects 90 per cent of the light that hits it, which helps keep it cool, and means that the ISS is easily visible in the night sky. It passes overhead quickly, but is visible as a bright dot passing from the southwest to the northeast. In the daytime, the ISS is much harder to spot, but if you know where to look, and the ISS happens to be at its closest to Earth, it can be spotted with

▼ The International Space Station, a feat of human engineering, has been continuously occupied for more than twenty years. With its solar panels, it is 356 feet (109 meters) long.

the unaided eye. If we make our space-supercarrier out of aluminum, which is very reflective and also light, then the reflectivity is 91 per cent; quite a close match to the ISS.

## 2. Surface area

At around 1,093 feet long and 252 feet wide (333 by 77 meters), a supercarrier is among the largest aircraft carriers. An object of that size has an area of 275,436 square feet (25,641 square meters). This is fairly sizable, as the ISS, our current largest orbiting satellite, is by comparison a relatively trim 41,064 square feet (3,815 square meters).

While keeping a similar reflected percentage of light, this much larger size means that our carrier would be much brighter than the ISS. In fact, it would be 15.6 times brighter than Venus. In the night sky, this would be a brilliant point of light moving quickly overhead, much like the ISS. If you were looking up at the right time, it would be hard to miss.

## Verdict

In daytime, the Sun dominates the sky, making it much harder to see anything beyond our own atmosphere. It's not impossible, though, as the Moon proves by being easily visible in the daytime sky. Our supercarrier wouldn't be as bright as a quarter Moon in the daytime sky, but it would be much brighter than Venus, which is already visible in the daytime. So our supercarrier would definitely be visible in the daytime, but it would be harder to spot than during the nighttime passes.

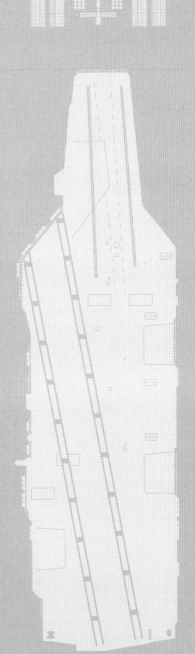

▶ The ISS, seen in silhouette at the top of the page, compared to the size of the largest supercarriers. The surface area of the ISS is dominated by its solar panels; our water-based craft is much larger, but hosts many more people.

# HOW VISIBLE IS LIFE FROM SPACE?

While floating in space, it's relatively easy to spot human activity on our planet. The easiest way, of course, is to look for the lights of our cities at night. The astronauts on board the International Space Station have taken some beautiful images of this, with the grid pattern of the roads in U.S. cities contrasted with the more tangled street patterns of older cities in other parts of the world.

If we were to go hunting for life on another planet, modifications of that planet's surface would be a big thing to look for, provided we had access to images of a high enough resolution. (This would be an "orbiting the planet" hunt, and not a "searching from here on Earth" hunt.) On Earth, we humans have not only illuminated the night, but have also carved into the fields, adjusted coastlines, and irrigated dry terrain.

But science fiction likes to take a different approach. Instead of searching for cities by their lighting, there's a vaguely phrased "signs of life" method, which often translates to hunting for signatures of warm bodies on the planet. If we turn this lens onto ourselves, we're certainly visible through a heat-based search, but only from a position close enough to the planet.

How many of the human-induced changes to our planet's surface we'd be able to see from space depends entirely on the resolution we can achieve with our camera. Resolution for an image depends on three things: proximity to the object in question, the wavelength of light being observed, and the number of wavelengths of that light that can fit across the telescope. If we're looking for warm bodies, a heat map would be needed. That means looking in the infrared, from at least an orbital distance around the planet.

In general, as you would expect, in infrared the poles of our parent planet show up as cold, and the equatorial regions as much warmer, but at this resolution you can't see any real details. Cities don't show up here, let alone individual humans. This is due to the combination of the wavelength (the infrared is a longer wavelength than optical light, so the resolution drops), the distance the satellite is orbiting the planet (about 440 miles, or approximately 700 kilometers up), and the size of the collecting area of the satellite.

If we just want high resolution to capture the smallest details on the surface of our planet, the best bet is to bring a really large mirror and camera (increased collecting area = better resolution), or to swop over from infrared to the optical range, though clouds would become a problem with the latter option. Earth's cloud layer is not very thick, nor is it very hot, and it tends to move over time, so if we wait long enough we should be able to see what's underneath any given cloud sooner or later. However, if we're observing a planet more Venuslike in its permanent cloud cover, the optical range of light is not going to be our friend.

On Earth, however, it works fine; commercial satellites in orbit can now image Earth down to a resolution of about 1 foot (30 centimeters). (Or at least that's as good as various militaries will allow them to disclose; super high-resolution imagery of Earth's surface is also used for military reconnaissance.) With optical high-resolution data you can look for geometric patterns. Perfect circles, squares, rectangles, or triangles are unlikely to happen naturally, so if you spotted widespread rectangles on the surface of Earth, that would usually mean you'd found a well-planned city or a farm, indicating some kind of intelligence at work.

▲ **Top:** Phoenix, AZ, as photographed by an astronaut on board the International Space Station. Phoenix is clearly visible at night, along with the grid pattern of its road systems.

**Bottom:** Land temperature in southern Europe and on the North African coast on June 29, 2019. While mountain ranges are visible, the city structures are much less obvious.

## HEAT ISLANDS

You can spot cities via infrared heat measurements; unless located in a desert, dense cities tend to be warmer than their surrounding areas. This is partly because we've cut down all the trees to build the city; another reason is that we've paved it with heat-absorbing asphalt. If the city has a lot of trees planted, this city "heat island" is less obvious. The resolution of heat-island images is about 100 feet (30 meters), which is too large to detect individual people. This is partly because the size of the mirror on the satellite is only 16 inches (40 centimeters) across.

▲ Buffalo, NY, as seen by LANDSAT. The darker colors indicate cooler temperatures, while brighter yellows are hotter locations. As with nightscape photographs, these heat maps show the presence of cities.

▶ Circularly irrigated fields in Kansas photographed from space by the *Terra* spacecraft, launched in 1999.

## SIGNS OF EXTRATERRESTRIAL LIFE?

Of course, the farther away from the planet you are, the harder this is to do—it's not the sort of scanning you can do while cruising the galaxy at high speeds. To map the whole Earth at low resolution (800–3,200 feet [250–1,000 meters]), the satellite-mounted MODIS instrument, in orbit at about 450 miles (725 kilometers) above Earth's surface, takes two days. So it's possible to detect signs of life on a planet via heat-based images if we're looking for evidence of cities, but not if we're looking for individuals, and not if you don't want to spend a few days in orbit around the planet.

Still, if you built a very large-aperture, wide-angle telescope, and had it orbiting the planet in space, you just might be able to spot people outside.

If you had a telescope that was 1,640 feet (500 meters) in diameter, you'd get resolution 150 times better than that of the Hubble Space Telescope. Even in the infrared, we'd be able to detect individual human beings if we gathered that much light—though to tell if anything was moving, you'd have to take a series of images and play spot the difference. (A series of extremely short exposures would also keep all your images from becoming too blurred, unless you've parked the telescope in geostationary orbit.) If you had an inkling of where to point your dish—and weren't reliant on mapping the entire planet—the civilization we've dreamt that Earth becomes in the future may yet be able to spot intelligent life walking around on other planets.

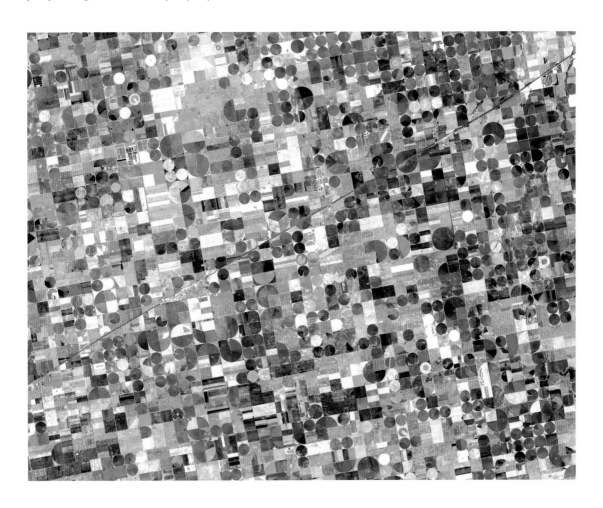

# Can you take a prehistoric photo?

It's an appealing idea to be able to see Earth as it was, many years past, but how might it be done? In theory, we could use the delay that comes with light's travel through the vastness of space to see our planet hundreds, even millions of years in the past. By putting a reflector in space, geometrically speaking, we could return sunlight to us that had bounced off Earth many years previously, thus getting the image of Earth from an earlier time. But how plausible is this, really?

**W**e do something like this already; the Apollo missions to the Moon set down mirrorlike reflectors on the Moon's surface, so we can bounce a laser beam off it and count how long it takes for the light to return to us. This provides some of the most accurate measurements to determine the distance to the Moon. In order to work out how we might take a prehistoric photo, there are a few steps to cover. First, we'll take a closer look at what was done on the Moon. Second, we'll think about the properties of light, and how much of Earth's reflected light we could realistically catch with our reflector. And finally, we'll consider what we'd be able to do with the light returned to us by our reflector in space.

▼ A typical mirror will never bounce light back the way it came unless you strike it at a 90-degree angle; any other configuration (such as the one here) will bounce the light away from its source. A retroreflector, by contrast, will always return light back the way it came.

▶ The Laser Guide Star (LGS) is launched from the Very Large Telescope's 26.9-foot (8.2-meter) Yepun Telescope and is aimed at the center of our galaxy. The laser beam creates an artificial star at an altitude of 56 miles (90 kilometers), which is used as a reference to correct images and spectra for the blurring effect of the atmosphere.

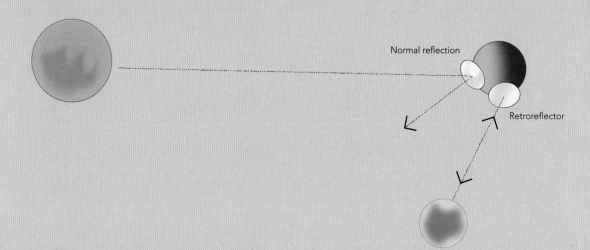

Normal reflection

Retroreflector

## 1. Return service requested

The reflector on the Moon isn't what we would normally consider a mirror; it's a retroreflector, whose main purpose in life is to bounce light directly back the way it came. A mirror doesn't do this, unless light strikes it at exactly 90 degrees. The retroreflector can solve the first of many problems: having the mirror exactly lined up with our telescope. Even a fractional shift in a standard mirror's position would be enough to bounce the light from Earth off into space, missing us entirely. However, with a retroreflector, we can successfully shine a powerful laser at the reflector, and have that light come straight back the way it came, regardless of how well aligned the reflector is.

If you'd like to bounce light off the Moon, the laser light will take around 2.5 seconds to travel the 478,000 or so miles (around 769,000 kilometers) there and back. This is a short trip, astronomically speaking, but we'll already start to see the next problem we'll have to tackle if we want to start seeing Earth reflected off a distant surface. We start to run out of photons.

## 2. Scattering the light

Even a laser, which starts out with all of its light focused into a very small beam, will spread out at larger and larger distances. Red lasers are more prone to doing this than green lasers, simply because the wavelength of light is shorter for green lasers, and this spreading is partially a function of wavelength. (Purple lasers would be even less prone to spreading out.) However, over several hundred thousand miles, even the highest wavelength lasers we can manufacture are going to spread out, and by the time this light gets to the Moon, the laser is only able to faintly illuminate the surface, and only a tiny fraction of that light is going to be bounced off the reflector and back to

Earth. The farther away you put your mirror, the worse this problem gets, because light will get increasingly spread out.

Earth does reflect sunlight out into space, so we're not in trouble there. However, this light begins to spread just as the laser light does, but since it's coming from a larger area to start with, it's never as compact and focused as the laser light. By the time we reach any astrophysically significant distance from Earth, this reflected "Earthshine" is very dim indeed. We can see Earthshine ourselves if the Moon is up at night. The reason the dark part of the Moon isn't 100 per cent black is that it's getting some reflected light from Earth. By the time this Earthshine light travels for several hundred light-years, you can well imagine that it has become very diffuse. And then, of course, it would have to travel several hundred light-years back, becoming even more diffuse on the return journey. We'd also have to conveniently find a mirror out in space with a clear light path between us and it—and if we just put it out there now, we'd have to wait a few hundred years for anything to come back our way.

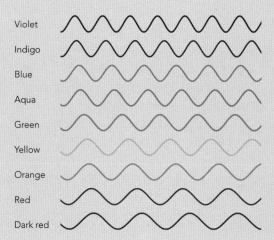

Violet
Indigo
Blue
Aqua
Green
Yellow
Orange
Red
Dark red

▲ The wavelength (the distance between the peaks and troughs) of light shortens the more blue the light is. The redder the light, the longer the wavelength.

▼ Earthrise, photographed in 1968 by the Apollo 8 crew, while they were in orbit around the Moon.

### 3. Identifying what we get back

Let's say we managed to get a few photons back from our several hundred light-years-distant reflector, which we can arbitrarily make sufficiently enormous that this would happen—would we be able to identify them?

Part of the reason we like using lasers for our Moon experiment is because they are all a very particular color of light, so we can count up the returning photons at that color, relative to photons of any other color, which we know are unrelated to our experiment. Earth is not a single color, and the atmosphere is incredibly complicated, so the set of photons that we would reflect would be a much more complex set than the laser beam we're firing at the Moon.

On top of this, when Earth is showing the most reflected light, it's because the angle between us and the Sun is the smallest. So when the Earthshine is the brightest, we're also most likely to be blinding our reflector with light from the Sun. The Sun is really, really bright. Stars in general tend to be a big problem for taking direct pictures of planets around other stars, because they're so luminous that they swamp out any of the reflected light from a planet, and we have to get really clever with how we block out the light from the star without blocking out anything else.

### Verdict

It seems unlikely we'll be able to conquer this task, even with a conveniently placed retroreflector. To make sense of our parent planet's past we make do instead with the stories Earth can tell us based on other measurements—without photographs we rely on geology, planetary science, and theoretical models.

▼ This composite image shows the exoplanet 2M1207b (the red spot on the lower left), orbiting the brown dwarf 2M1207 (center). The photo is based on three near-infrared exposures made by the NACO adaptive-optics facility at the VLT Yepun telescope (shown in the image on page 29) at the ESO Paranal Observatory.

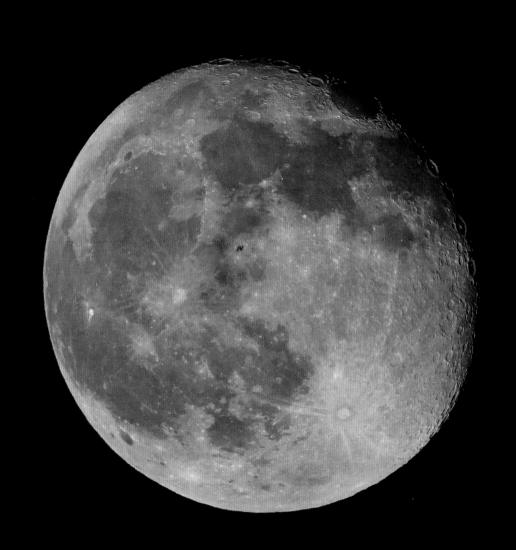

2

# THE
# MOON

The Moon is a constant presence in our skies. Its harsh, jagged surface is the only terrain outside of Earth that humans have trodden, and it has been the source of much curiosity about the cosmos. Let's investigate what we know of our lunar companion and how it affects us here on Earth.

# WHY DO WE ONLY EVER SEE ONE SIDE OF THE MOON?

The Moon is simultaneously a constant and ever-changing presence in our skies. As the Moon moves in its orbit around our planet, and the angle between it and our Sun changes, the illuminated portion of the Moon is perpetually creeping along its surface. And yet, if you know where and when to look, it is always a feature of both our daytime and nighttime skies.

According to current theory, the Moon was formed when a large object, about the size of Mars, slammed into the early Earth a few tens of millions of years after the solar system began (still about 4.5 billion years ago). The impact shredded Earth's outer layers and destroyed the other object. The debris from this astoundingly violent encounter would have formed at least a ring, and possibly a haze of very hot debris around our planet. As that debris cooled, it collected back into the object we now know as the Moon. As a result of this collision, the Moon is built out of the same rock as much of Earth, and it is a much larger moon than you might expect a planet the size of Earth to have. The Moon has been our companion since that early collision, and our nearest and most visible reminder of the expanse of space beyond our planet.

The constancy of the Moon's companionship is reflected in another way—to our eyes on the surface of Earth. We always see the same side of the Moon. The same, familiar features of the lunar landscape face our little planet, no matter what time of day or phase of the Moon, nor whether there is a solar eclipse. The features we see at the full Moon, where the Earth-facing side of the Moon is fully illuminated, are the only ones we'll ever see from here. A half Moon appears when only half of the near side of the Moon is illuminated, with the other half plunged into shadow. The unobservable far side of the Moon is also half illuminated, half shadowed.

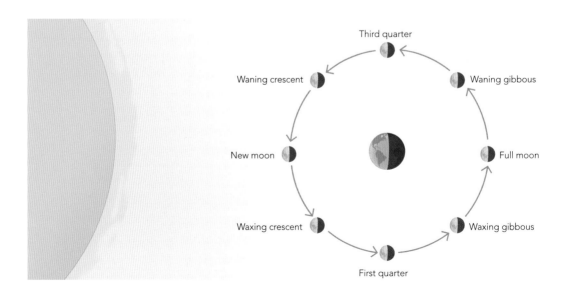

Third quarter

Waning crescent

Waning gibbous

New moon

Full moon

Waxing crescent

Waxing gibbous

First quarter

◀ The sunward-facing side of the Moon is always illuminated, but the fraction of that illuminated face that is visible from the surface of Earth changes depending on where the Moon is in its orbit around Earth.

▲ Daedalus crater, on the far side of the Moon, in the center of this image, is about 50 miles (around 80 kilometers) across. This image was taken by the astronaut Michael Collins in orbit around the Moon during the Apollo 11 mission.

## THE FAR SIDE

The far side of the Moon is an almost impossibly foreign-looking place. It's entirely full of craters—unlike the near side, where the dark patches of the Moon (known as "mare") look relatively clear of craters, the far side of the Moon has no uncratered space left.

The reason we never manage to see this "other side" of the Moon is because of a trick of the Moon's orbit around our planet. The Moon has settled into a rotation and orbit that match each other—for every trip around Earth, the Moon rotates once. One full "day" of the Moon takes about four weeks—the same length of time it takes to complete a cycle of the phases of the Moon. If the Moon didn't rotate, we would see all sides of it, including its strange cratered far side. The only way for us to see such a constant face of the Moon is if it's also very slowly rotating.

▼ **To rotate or not to rotate**
   The far side of the Moon, though just as illuminated by sunlight as the near side, is held permanently away from Earth's surface because the rotation speed is the same as its orbit. If the Moon did not rotate at all, we would see all sides of the Moon, depending on the phase.

# If Earth were tidally locked to the Moon, the Moon would never rise or set in our skies, and half the planet would never see it at all.

Think about sitting in a spinning chair, and holding an open book at arm's length. If you hold the book and spin yourself around, you will always be able to read the text within the book. But the book's orientation with respect to the room has done a full circle as you spun. Now imagine you asked a friend to hold the open book and walk around you, always facing the north wall. After half a spin, you'd be looking at the spine of the book, not the inside text.

The configuration of rotating to match an orbit, the way the Moon does, has been dubbed "tidal locking," and is simply defined as a setup where one face of an object is continually facing another as it orbits. Sometimes both objects can be tidally locked to each other, which isn't the case for Earth. If Earth were tidally locked to the Moon, the Moon would never rise or set in our skies, and half the

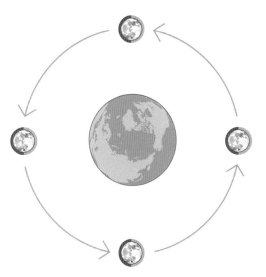

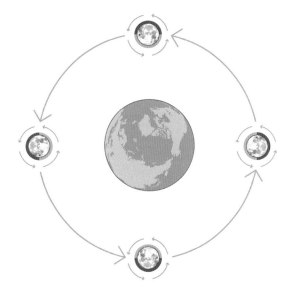

▲ The far side of the Moon, illuminated by the Sun, as it crosses between the DSCOVR satellite's Earth Polychromatic Imaging Camera (EPIC) camera and telescope, and Earth—1 million miles (1.6 million kilometers) away.

planet would never see it at all. But this is the situation for the dwarf planet Pluto and its largest moon Charon—both are tidally locked to each other.

The amount of time it takes to orbit around the planet will vary from object to object (Phobos, one of the moons of Mars, is tidally locked and orbits Mars every 8 hours—way faster than our moon), but as long as the object is tidally locked, the length of the "day" on that moon will match the length of time it takes to go around the planet it orbits.

Tidal locking is a remarkably stable setup. Now that the Moon has got into that configuration, it will remain there for a long time. The reason why it's so stable is buried in the physics of tidal forces.

THE MOON

# HOW DO THE TIDES AFFECT EARTH?

For those who live by the sea, you may already know that it is the Moon that creates the ocean tides. High and low tides are the ocean's response to the gravitational pull of our Moon. A glance to the sky can therefore tell you if the tide is high or low. If the Moon is overhead, you're at high tide; if it's at the horizons, your shores are at low tide.

We have ocean tides because the Moon is massive and relatively close by, and so the gravitational force from the Moon on Earth is significant. In astronomical terms, if the strength of a gravitational pull changes dramatically across a specific object, that's a tidal force. The Moon's gravitational pull is the prototype for this. The pull of the Moon on the near side of Earth is significantly stronger than its pull on the far side, because Earth and the Moon are so close together. There are no limitations on what kinds of objects can experience a tidal force. A human could feel a tidal force if you went too close to a black hole; a comet can feel a tidal force if it goes too close to Jupiter; and a moon can feel a tidal force because of its planet.

On Earth, the presence of the Moon means that the side of the planet facing the Moon at that moment feels a particularly strong gravitational pull, and the exact opposite side of Earth feels a particularly weak pull. If Earth were softer, this would stretch it out into more of an elongated shape, toward the Moon. However, Earth is made of rock, and this material isn't easily stretched.

The rocky material just has to deal with a certain amount of strain. However, the water coating our planet is shifted around easily, and it is free to rearrange itself in response to a stronger pull on one side of the planet, or a relatively weaker pull at the opposite side. So this stronger pull on the Moon-facing side results in a high tide underneath the Moon. Paradoxically, high tide also exists at the weakest gravitational pull—the exact opposite side of Earth. This weakest gravitational pull allows the water to drift away from the Moon more than it would be able to on the Moon-facing side, and so the water there can hang back, relative to the rocky sphere of Earth, which pretty much stays put, thus forming another high tide. Low tide, meanwhile, occurs where the oceans are being stretched from, exactly in between the two high tides.

▼ The difference between high and low tide can be particularly stark in certain parts of the world. Here, the ground at Hopewell Rocks in New Brunswick, Canada, is exposed at low tide, but submerged at high tide.

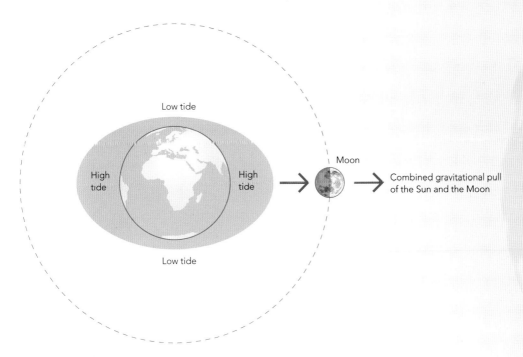

High
tide

Low tide

High
tide

Low tide

Moon

Combined gravitational pull
of the Sun and the Moon

▲ **Tidal forces**

The Sun and the Moon can both influence the
tides; when they are aligned (as shown here),
the tides will be particularly high and particularly
low. However, since the Moon is much closer to
us (not drawn to scale here), the influence of the
Moon is generally much greater than that of the
Sun, even though the Sun is much more massive.

## SLOWING THE MOON'S SPIN

The gravitational tidal forces that distort the
oceans can also resist the rotation of an object.
When the Moon was very young, it would have
rotated at a much faster speed, and probably
would have orbited Earth at a different speed.
The Moon was almost certainly not tidally locked
when it first formed. Had any observer been on
that early Earth, they could have seen all sides
of the Moon as it spun.

The gravitational pull from Earth—which,
like the tides due to the Moon, pulls on the side
of the Moon closest to Earth more than the far
side—resisted this faster rotation. This resistance
due to the gravitational pull of Earth gradually
slowed down the spin of the Moon until it was
no longer rotating faster than it was orbiting.
Once the Moon's rotation had slowed so much
that a single face was always facing the surface
of Earth, that was the beginning of its tidally
locked orbit, and the Moon has stayed
in this configuration ever since.

The Moon also has the same influence on
Earth, but since the Moon is so much less
massive than Earth, this resistance to rotation
takes a much longer time to impact on Earth's
spin. However, it's still a measurable effect! The
Moon is slowing down the rotation of Earth by
about 15 microseconds every year, gradually
lengthening our days.

## ATMOSPHERIC TIDES

The air in our planet is also relatively free to move around—as a gas, it doesn't have the same pressure to remain fixed in place like the rocky underbelly of our planet. Our atmosphere also has regular swellings and subsidings, much like the ocean tides, but it's the Sun that triggers these changes, instead of the Moon. While we also call these changes to our atmosphere "tides," it's in the older sense of the word (a regular change), and it's not the Sun's gravity that's making such a large impact.

You can guess that the Sun's gravitational impact on atmospheric tides must be small by looking at the ocean tides. If the Sun were a major player in ocean tides, then high tide would happen at local noon every day, instead of varying based on the Moon's position in the sky. The Sun is still a player—the highest tides occur when the Sun and the Moon align so both of them are pulling in the same direction—but the difference between a normal high tide and high tide when all forces align isn't as large as the difference between high tide and low tide; the Moon wins that round.

The key difference between atmospheric and ocean tides is that, unlike our liquid oceans, the atmospheric gases have some extra things that they rely on. Liquids are fairly straightforward under normal conditions; they fill their vessels and are relatively constant in density—it's hard to compress or expand a liquid by very much. Gases, on the other hand, are highly variable in density. It's fairly easy to compress a gas—we do this by talking. Gases are also highly sensitive to temperature; one of the easiest ways to make a gas more dense is to simply cool it down. Conversely, if you want to make a gas puff up and take up more space, heat is an easy way. Each particle of gas gains a little more energy than it had before, and it bounces around a little faster, so the whole thing becomes slightly more puffed up. So the Sun, as an enormous source of heat, has a pretty major influence on the gas in our atmosphere by simply heating it up.

▼ Sunrise over Earth, taken by the Expedition 40 crew aboard the International Space Station.

When the Sun is shining on the atmosphere (in other words, anytime it's daytime), the Sun heats up the gas that surrounds our planet and makes the whole volume expand outward toward space. This expansion can be measured at sea level as a very slight reduction in atmospheric pressure, but it is a lot more dramatic at higher altitudes, where the density of gas is really low to start with—as the atmosphere underneath it expands, suddenly there's an upwelling of denser gas from below. This particular effect isn't a gravitational tidal force at work, but the official terminology is an "atmospheric tide," and it is a regular, cyclical change.

Of course, the Moon does have a gravitational tidal effect upon the atmosphere, but like the Sun's impact on our ocean tides, it's a much weaker effect than the heating provided by the Sun. If the Moon were the main cause behind this atmospheric stretching, it would work the same way as the ocean tides. High tide would mean that you also had the most atmosphere above you, instead of what we see: a 24-hour cycle of our atmosphere heating and cooling under the Sun's rays.

▼ **Terrestrial atmospherics**
The heat from the Sun expands the gas that makes up the atmosphere. At night, by contrast, the gas cools and shrinks back down.

Small expansions in each layer of the atmosphere cause a cumulatively larger expansion of the atmosphere overall.

Sunlight provides the heat required to expand the atmosphere.

The night side of the planet allows the atmosphere to cool away from the light of the Sun.

# WHAT COULD CAUSE THE MOON TO CHANGE SIZE IN THE SKY?

Not only is the Moon tidally locked to our parent planet, but its orbit happens to be almost exactly circular, which means its distance from us doesn't vary that much over the months. Because the Moon isn't going on long, looping journeys away from Earth, our view of the Moon is of an object that is very close to a constant size in the sky.

Without something catastrophic happening to the Moon, like a giant impact, we wouldn't expect its physical size to change. Because it is made of rock, like Earth, it won't evaporate over time, even in the harsh environments of outer space. Radiation from the Sun does bleach all the surface material on the Moon, but it shouldn't remove any significant amount of material. If something catastrophic did happen to the Moon, it would certainly fling material away from its surface. However, large impacts are increasingly rare in our solar system—we effectively ran out of all the chaotic large objects in impacts much earlier in the lifetime of the solar system. This means that a very large-scale event is extremely unlikely (although not impossible).

There can be minor changes to the Moon's size in the sky, though. The Moon is on a slightly elliptical orbit, which means it has a closest and farthest distance from Earth as it orbits. The difference between the closest and farthest approach is about 26,000 miles (roughly 42,000 kilometers). Considering that, on average, the Moon sits 238,856 miles (384,402 kilometers) away, this is a relatively minor shift in distance. When the Moon is closest to us in its orbit, and also full, you get a "Supermoon," but as you can see from the images opposite, it's not that big a change from the farthest and smallest the Moon ever gets from us (the "Micromoon")—the size difference is about 12 per cent.

> Because the Moon is broadly interacting with Earth gravitationally, the Moon and Earth are slowly exchanging energy.

On top of this orbital change, because the Moon is broadly interacting with Earth gravitationally, the Moon and Earth are slowly exchanging energy. As we saw, this interaction is responsible for the slowing of Earth's rotation, and additionally allows the Moon to drift away from Earth. This happens at an incredibly slow rate: we're talking about 1.5 inches (40 millimeters) every year. Compared to the normal orbital variation of 26,000 miles, this effect is barely noticeable, and certainly won't be to the human eye.

But the next time you hear something in the news about a Supermoon, remember that it's the closest the Moon will ever be to Earth, if only by a few millimeters. By the same token, each Micromoon gets a few millimeters more distant, and a tiny bit smaller, each year.

▲ The Moon over Washington D.C. on December 3, 2017. Taken with a telephoto lens, the Moon appears large because of the extreme zoom used.

▼ The actual distance between the Moon and Earth varies over the course of the Moon's orbit. This image compares the Moon's apparent size when it is closest to and farthest from Earth.

# WHY DOES THE MOON HAVE A HIGH-ENERGY GLOW?

While large objects are relatively rare in the inner solar system, small objects abound. Without a protective atmosphere like Earth's, which serves to burn up the smallest pieces of grit around our planet, the Moon does not get to enjoy meteor showers. Any meteoroids or other objects on a collision course with the Moon can go crashing straight into its surface, with no slowing whatsoever. (There's a whimsical term for this kind of impact: "Lithobraking"—to slow down by hitting rock.) Usually, this ends badly for the fast object.

Inside of our protective parental shell, it's easy to forget how harsh the solar system can be, and one of the reasons it's so harsh is a constant flow of cosmic rays, usually coming into our solar system from elsewhere in the galaxy. Cosmic rays are actually highly accelerated tiny pieces of grit and not rays of light at all. Many of them are subatomic particles in the form of protons, constantly streaming throughout the universe. Whenever they smash into the surface of the Moon they release so much energy that they can produce an extremely high-energy form of light: gamma rays.

Gamma rays are one of the highest-energy forms of light in the universe—their wavelength is around a picometer (that's one trillionth of a meter, or 0.000000000001 meters), which is roughly the same order of magnitude as the size of a hydrogen atom. These wavelengths are a thousand times shorter than visible light, and therefore invisible to our eyes.

## DANGEROUS RAYS

The high energy of gamma rays means that this form of light is extremely damaging to life. Ultraviolet (UV) light, which is also too high-energy for our eyes to absorb, but considerably less energetic than gamma radiation, is already damaging to the human body. UV light doesn't penetrate far into your skin—it can damage the surface layers, giving you a sunburn (and increasing your risk of skin cancer), but it won't damage anything past the first few layers of skin. Gamma rays can penetrate much farther into your body, and can destroy or alter the DNA within your cells, causing drastic changes to the replication instructions of those cells. This can cause radiation sickness and cancer to appear, but without any surface burning of your skin. Very fortunately, Earth's atmosphere is opaque to gamma rays, and it does a wonderful job of protecting us from the beating our cells would otherwise take. Gamma rays that encounter our atmosphere will very rapidly run into an atom, smashing into it hard enough to create an electron and a positron. The electron and positron both lose a tiny bit of energy, and they radiate away the extra energy as a slightly lower energy gamma ray. This process repeats until the high-energy light has been depleted of enough energy that it's no longer counted as a gamma ray.

On the other hand, the fact that our atmosphere is such an effective wall to this wavelength of light means that we can't directly observe any of the gamma rays produced elsewhere in the universe, or on the Moon, from the surface of Earth. In order to look at this light, we have to use a space telescope, which avoids the interference of the atmosphere. At the moment, the Fermi Gamma Ray Space Telescope is observing the whole sky for gamma ray sources, which can tell us about our own galaxy, other galaxies, and the explosions of massive stars in supernovae.

◄ Blazar 3C279 is one of the brightest sources of gamma rays in the sky. Powered by a supermassive black hole (see page 192), these gamma rays are produced by tremendously hot gas swirling around a black hole.

▼ **Gamma rays in Earth's atmosphere**

An incoming gamma ray will interact with particles in the upper atmosphere, which causes a cascade of less energetic particles to run into other atmospheric particles, until their energy is depleted. Light is produced as part of this process, which can be detected by telescopes on the ground.

Because the Moon is constantly being bombarded by cosmic rays, this means that it is also continually producing gamma rays as the cosmic rays come to a screeching halt. To a space-based observatory, this means that the Moon glows in gamma rays. The Moon, as viewed by a gamma ray telescope, is actually brighter than the Sun—which, very fortunately, does not produce much in the way of gamma radiation. If it did, the gamma radiation from such a close, bright source might have destroyed our atmosphere, removing our shelter from this cosmic hazard.

This kind of cosmic pummeling is something we will have to guard against for any kind of future travels outside our home world, and one of many reasons why underground structures on other planets have been suggested as better places to put our astronauts than the surface—that way the ground can take the hits instead.

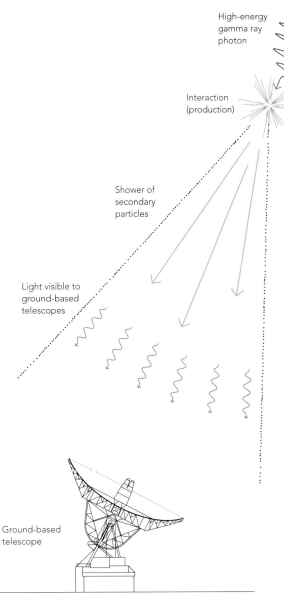

High-energy gamma ray photon

Interaction (production)

Shower of secondary particles

Light visible to ground-based telescopes

Ground-based telescope

# How much of a mess will you make if you have a door that opens to the lunar surface?

Earth and the Moon are tremendously different environments, and perhaps nothing illustrates the differences between our parent planet and its companion more than the thought experiment of opening a door between the two worlds, and allowing humans to step between them. Between the vast differences in temperature (the Moon's daytime temperature is 224°F/106°C), and the complete lack of an atmosphere on the Moon, opening a doorway between the two is not as simple as going outside on a hot day.

We have a couple of important things to consider. The first step is to look at how the lack of an atmosphere on the Moon would react with the standard air pressure we experience here on Earth. Secondly, we'll need to consider how we might counter these reactions (spoiler: it isn't easy). Finally, there's the small matter of figuring out how not to be flung out into space.

## 1. One wind, coming right up

The most immediate issue facing us as we push open that door is the lack of air on the Moon; humans need this pretty reliably to continue living. However, if we were to open a door between a space that has a lot of air, and a space that has none, the air pressure of our atmosphere on Earth will immediately begin to flood through the doorway, out into the vacuum of space on the lunar surface. This will produce a tremendously strong wind.

On the one hand, you're not likely to run out of air, with Earth's atmosphere blowing at you like air out of a balloon. This will also protect you somewhat from the heat of the daytime lunar surface, since all the wind chill equations I could find indicate that room temperature air (starting at 68°F, or 20°C) will only get down to 59°F (15°C) or so, no matter how fast it's rushing out of a doorway. If you

happened to open your door onto lunar night, you'd probably get instant snow, as the water vapor in the air freezes.

In this set-up, the lack of air isn't an issue. The problem is that you have too much of it, given exactly how fast the air would be rushing out of the door. There's a handy equation called the Ensewiler formula to convert a pressure difference into a wind speed, which is written out as $P = 0.002496\ v^2$. P is the pressure difference in pounds per square foot, and the velocity comes out in miles per hour. An opening between Earth and the Moon is a door between standard air pressure (one atmosphere, or 2,116.216 pounds per square foot) and no atmospheric pressure at all. So if we put a difference of one atmosphere into the Ensewiler equation, we get a velocity of 920 mph (1,480 kph), which is kind of a problem. This is 1,349 ft/s, and faster than the speed of sound. For some context, the fastest recorded wind speeds on Earth are 253 mph, or 407 kph (recorded in Tropical Cyclone Olivia, which hit Australia in 1996) and an F5 tornado that hit Oklahoma in 1999, which clocked in at 302 mph (486 kph). Felix Baumgartner's jump from the edge of space in 2012 got him up to a speed of 843 mph (1,357 kph) before he pulled his parachute, but he was wearing a massive protective

suit, specifically designed to keep him safe, and he crossed the sound barrier quite high in the atmosphere, where the air is not very dense. Unprotected, the shock from entering a 920-mph wind won't kill a human immediately—a shock wave from a bomb only becomes lethal at about 1,500 mph (around 2,400 kph). It's still not doing anyone any great favors, but you won't die from the pressure impact.

## 2. What doesn't kill you may fling you into space

The biggest challenge would be how to restrain yourself on the surface of Earth, without being flung headlong into space by the sheer wind pressure. If you tried to keep yourself in place by fixing heavy-duty handles to the Earth-facing side of the doorway and holding yourself against the wind, you'd rather rapidly run into some problems. The average person has grip strength of approximately

▲ Apollo 11 astronaut Buzz Aldrin is photographed on the lunar surface with the landing module behind him. Carrying their own supplies of oxygen, and heavily insulated from the blazing Sun, the Apollo astronauts' space suits were bulky but necessary to protect them from the harsh environment on the Moon.

500 Newtons of force. The drag force from 920 mph against a human body is 15,154 Newtons in the other direction. Five hundred Newtons is not going to be enough to keep hold of anything against that pressure. 500 Newtons will keep you holding on to something in the face of 182 mph (293 kph) winds, but nothing more than that. No matter how much a person clung on, the force of the wind would blow them away.

This calculation assumes that our human daredevil is able to continuously grip the handle until the wind speed overcomes their grip strength. In actuality, being dragged through such an inter-world doorway would be very much like being thrown out of a jet traveling at 900 mph (1,450 kph), and told to hang on to a trapeze bar.

The first thing that would happen is that both of your shoulders would dislocate. Shoulders dislocate with 325 Newtons of force, well below the force the wind is exerting. If your shoulder is dislocated, there's a very low probability that you're going to be able to hold on to much of anything, since the major nerves traveling down the arm are being stretched a lot more than they usually are, and nerves don't like being stretched. You could rig yourself so that you were attached via steel cabling to something anchored on Earth. A 1.2 x 1.2-inch (3 x 3-centimeter) cable (which is pretty industrial) will hold you fast, but you'd definitely need the assistance, because the wind would be faster than the terminal velocity for a human (the fastest speed a human will reach falling through the air), so it would certainly carry you along with it.

You'd also want to make sure that the Earth-side room had nothing at all that could move when subjected to a 920-mph wind. At those speeds, a 10-pound (4.5-kilogram) object hitting our human would inflict about 1,860 Newtons of force, which would break a finger bone. A 5-pound (2.3-kilogram) object would cause a wrist fracture similar to the kind of injuries that boxers can get, regardless of how well attached you are.

### 3. Lithobraking

With such a wind at your back, it's tempting to think that you would be given enough of a push to propel you away from the surface of the Moon, leaving you drifting forever in the space between the planets. Anything that gets blown through the door between Earth and the Moon would have an interesting journey, but even traveling at 920 mph isn't enough to escape the gravitational pull of the Moon. Escape velocity for the Moon is 5,324 mph (8,568 kph), meaning that 920 mph is six times too slow to escape. We can, however, calculate how long any object would spend flying in space before crashing down to the surface of the Moon. In an ideal case where the door lies on the ground, our hapless traveler would be flung out straight "up" from the surface of the Moon. If we assume that any objects were sped up to wind speed, they'd have eight and a half minutes of flight (reaching a grand old height of 32 miles [52 kilometers] above the surface of the Moon) before crash-landing on the Moon again at 920 mph (probably destructively).

### BRACE FOR IMPACT

There is a story about a fighter jet pilot, Captain Brian Udell, who ejected from his aircraft at roughly 800 mph (around 1,300 kph). His shoulder and knee were dislocated, an ankle was broken, and all the capillaries in his face burst from the force of the wind—his head swelled up to the size of a basketball, and his lips swelled to the point where he had a hard time moving them. He survived, but he was in hospital and undergoing physical therapy for six months afterward.

The wind between Earth and the Moon would be about 110 mph (around 180 kph) stronger than that, so it's probably safe to assume that the capillary breakage suffered by Capt. Udell would also pose a problem for any human standing in the way of this inter-world wormhole, whether or not they were well attached to the ground. This is not something to try without really serious protective measures.

## Verdict

You will make an incredible mess if you open a door to the Moon. While it wouldn't kill you instantly, the wind speed would do quite a lot of damage to an unprotected human, and anything or anyone that lost its grip would have less than 10 minutes to admire the view of the Moon before slamming back into the lunar surface at extremely high speeds.

## ▼ Escape velocity

The escape velocity for any object depends on both how close it is to the massive object, and also how massive the object is. The closer to the surface, the faster the object will have to move, in order to fully escape the gravitational pull of the massive object. If you surpass the escape velocity, you'll leave with energy to spare; if not, you'll fall back to the surface.

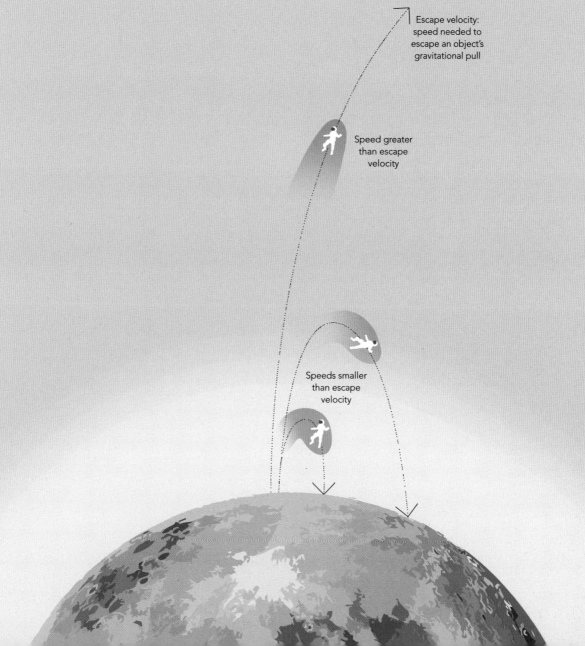

Escape velocity: speed needed to escape an object's gravitational pull

Speed greater than escape velocity

Speeds smaller than escape velocity

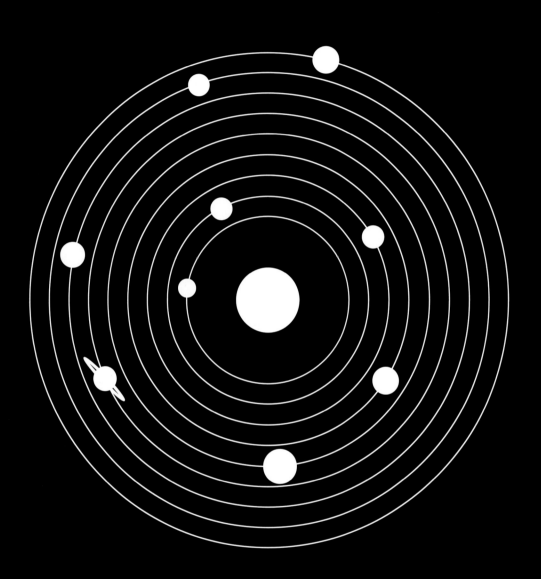

# THE SOLAR SYSTEM

The other seven major planets of the solar system represent Earth's cosmic siblings, all with commonalities and stark differences. From the gas giant planets in the outer solar system to the rocky worlds closer to the Sun, how does such a complex system come to be?

# HOW WAS OUR SOLAR SYSTEM FORMED?

Our solar system is made up of the Sun and all the planets, along with some other, smaller objects, formed roughly 4.5 billion years ago, out of a cloud of slightly rotating gas and dust. This cloud was just dense enough to start to collapse in on itself, just heavy enough to feel its own weight.

The eight major planets are Mercury, Venus, Earth, Mars, Jupiter, Saturn, Uranus, and Neptune. Mnemonics to remember the order abound, but the International Astronomical Union's current favorite is "My Very Educated Mother Just Served Us Nachos." We live on the third closest planet to the Sun, and we've designated the distance between our planet and our star as a unit of distance for the entire solar system: the astronomical unit (au), which is roughly 93 million miles, or about 150 million kilometers. Mercury, the closest planet, orbits the Sun at 0.4 au, four-tenths the distance of Earth from the Sun. Venus orbits around 0.7 au, and Mars at 1.5 au. The outer solar system, beyond Mars, orbits much farther from our

Sun, ranging from 5.2 au for Jupiter to a little under 10 au for Saturn, and just over 19 au for Uranus. Rounding off the major planets is Neptune at a distance of 30.1 au from the Sun—30 times more distant from the Sun than our own Earth.

In between and beyond our eight major planets there are other objects: comets, asteroids, and minor planets. The minor planet class contains things in our asteroid belt, between Mars and Jupiter, like Ceres, but also the rather beloved Kuiper Belt object Pluto. Let's explore our Earth's planetary siblings before we go any farther up our cosmic family tree.

**▼ Planetary orbits**

The orbits of all the major planets around the Sun are illustrated here with dotted lines. They are all nearly circular and are oriented so that they lie in a single flat plane. Distances are not to scale.

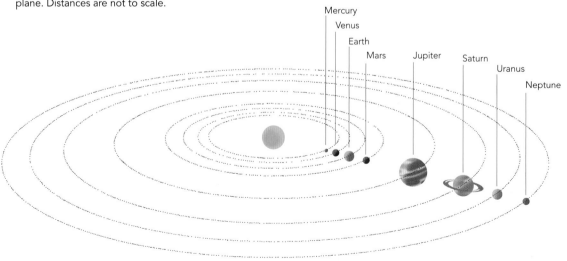

Mercury
Venus
Earth
Mars
Jupiter
Saturn
Uranus
Neptune

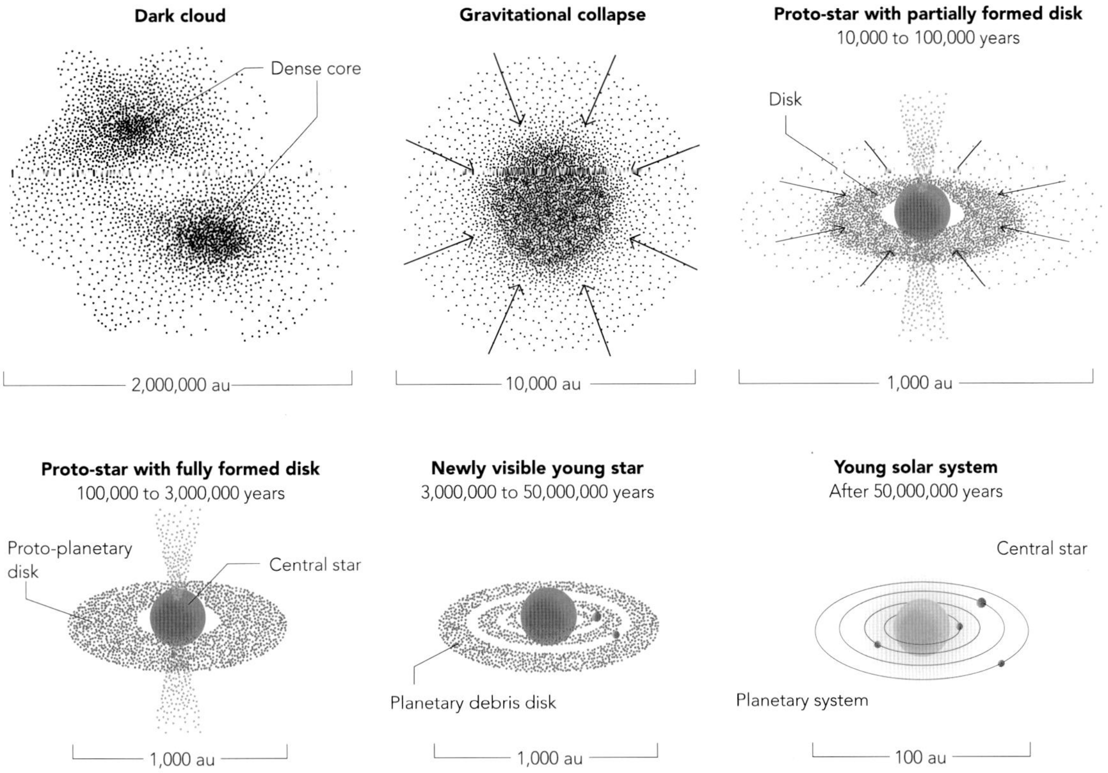

**Dark cloud**

Dense core

2,000,000 au

**Gravitational collapse**

10,000 au

**Proto-star with partially formed disk**
10,000 to 100,000 years

Disk

1,000 au

**Proto-star with fully formed disk**
100,000 to 3,000,000 years

Proto-planetary
disk

Central star

1,000 au

**Newly visible young star**
3,000,000 to 50,000,000 years

Planetary debris disk

1,000 au

**Young solar system**
After 50,000,000 years

Central star

Planetary system

100 au

## THE STORY OF THE PLANETS

The story of the origin of our planets is closely tied to the origin of the Sun itself, which sits at the very center of the solar system—before the formation of the Sun, the planets were nothing more substantial than dust and gas. The vast majority of the cloud that produced the Sun sank into the rapidly collapsing center, forming the nucleus of our solar system. Around 100 million years later, the planets began to form around the young Sun. This collapse wasn't perfect, or we'd have wound up with solitary stars and nothing left over to form planets, instead of a rich family of planetary companions for our Earth.

To get from a diffuse cloud of gas to the planets, a few things have to take place. To get the planets as we know them, we first have to make a star, and to make a star, we have to get gravity to pull a gigantic cloud of gas down on itself. This might begin by having the gas cloud disturbed, jostled

▲ **Building a solar system**
The process of building up a solar system is thought to take a long time—some 50 million years. An initially nebulous cloud of gas, with no particular structure, cools and collapses under its own gravity. Given enough time, enough material in the center will coalesce into a proto-star, with a disk of material surrounding it. Within the disk, planets may form, and eventually the proto-star will begin fusing elements, which signals its graduation into a full star.

into a denser state than it had been. Another possibility is that the cloud could simply have had enough time to cool, and with cooler temperatures come higher densities. This is the same principle behind hot air rising over cold air, or cold air sinking below hot air—the cold air is denser, and sinks below the less dense hot air.

▲ This artist's illustration shows a young sunlike star encircled by its planet-forming disk of gas and dust. Streams of material are seen spiraling from the disk onto the star, increasing its mass.

## GRAVITY-ENHANCED

Once the cloud begins to collapse downward, gravity joins in, and the cloud will also begin to collapse inward because of its own weight. As it does so, the cloud will also start to rotate a little. The more the cloud collapses, the more it rotates, until instead of a blobby cloud of gas, you have a relatively thin, rotating disk of gas. You can test this at home if you have a chair that spins. If you start yourself turning a little bit with your arms and legs extended, and then pull your legs and arms in quickly, you'll find yourself spinning much more rapidly than you were when you started out. Pulling your arms in plays the same role that gravity played in the collapse of the early solar system.

Once we've got this far, our cloud of gas has converted itself into a spinning disk of gases, but most of the gas has gone straight down into the middle. If there's enough there, that central core will be able to form into a star. But even with the

### RELATIVE HEAT

Relative to the impressive heat of the proto-star, which hasn't quite reached the pressure and temperature to light up with nuclear fusion, its disk is somewhat cooler. The proto-star clocks in at 2,000–3,000 degrees Kelvin (K; see page 203), or 3,140–4,940°F (1,727–2,727°C), while the nearby surrounding disk is "only" 500–1000 K (440–1,340°F , or 227–727°C), though the disk will drop down to temperatures below the freezing point of water farther away from the forming star.

star absorbing the vast majority of the material that collapsed, there's still a reasonable amount of gas that didn't make it in that far and is hanging around in the disk. This material didn't go into the star because it was too far away to be easily pulled inward, or was spinning too fast. This leftover gas and dust out farther from the star is still relatively cold (see the box opposite), and cold gas tends to collapse. (This is how we got the disk in the first place!) There's usually not enough mass in the rest of the disk to form another star, but the gas and the dust can begin to condense down and stick together, gradually forming small grainy bits of stuff—little chunks of solid material, in among the swirling gas around the star.

These new chunks of stuff orbiting around the star will crash into each other (see the box on the right) until the proto-planets gain enough mass to start attracting objects through gravitational forces, which can help them gain mass even more quickly. Once a particular lump reaches a certain mass, its own gravitational weight will start making it rounder and rounder. For some objects, this process will continue until they form a fully fledged planet.

## PLANETESIMALS

This ideal case of catching smaller lumps of material is rarely a smooth process. In the earliest stages of building up a planet, the speed with which it can gain mass is largely dependent on how much nearby matter there is to be wafted in the right direction to clump on to our object. Ultimately, this means that some infant planets will grow more rapidly than others, if they find themselves in a region that has a lot of material with which to grow. All of which means that the early solar system was a very hectic place, with swarms of "planetesimals" of various sizes. Some of these planetesimals will unavoidably have been absorbed into one of their larger neighbors, building up the largest planets to the sizes we see today.

### GROWING PAINS

When pieces of matter orbiting a star crash into each other, a numbers of things can happen. They can either knock each other apart (meaning the growing process has to start over), bounce off of each other (meaning that the encounter did neither of them any good), or they'll glue themselves together and become one larger lump.

The gas disk of the early solar system also determined the direction the planets were going to orbit around the central star. None of them were going to suddenly come to a stop and reverse course without something very violently turning the whole thing around, which would be hard to arrange. The planets in a typical solar system should also be pretty close to orbiting in a single plane around their star—that plane effectively shows us where the gas disk once was. Our solar system is no exception to these expectations. Every planet goes around the Sun in the same direction, and the eight planets are all aligned in a very thin plane. We can see this ourselves in the night sky: all the planets always appear along a single arc in the sky, which, coincidentally, the Moon also follows, named the ecliptic.

# DO ALL PLANETS SPIN THE SAME WAY?

The same kind of spin-up that happened when the cloud of gas collapsed downward into a disk continues when you go from a disk of material to a set of planets. Planets are more dense again than the disk of the early solar system, so we expect most planets to continue the pattern of spinning. And indeed, in our solar system, which is our easiest to observe set of planets, all the major planets are spinning around their own internal axis.

We're well acquainted with the rotation of Earth, of course, because that's what creates our days. We all know it takes 24 hours for Earth to rotate, but 24 hours isn't the rule within our solar system. In fact, of all the other planets, only Mars rotates at a similar speed.

With two exceptions, every major planet in our solar system rotates in the same direction as Earth. If we had a bird's-eye view of our solar system, where we'd flown into space "up" via the North Pole and looked back down, most of the planets would be rotating counterclockwise—from the west toward the east.

This consistency in spin direction is expected, again because all the planets are formed from a disk of gas that was itself spinning. It would be a violent, difficult task to reverse that spin. But it's not impossible, and our solar system has a few examples that show it can happen.

Venus and Uranus spin in an unusual direction. Both planets should have formed with a rotation aligned with every other planet in the solar system, so to get to their current upside-down and sideways state, something must have changed them. In fact, that's probably what happened.

The early solar system didn't just form eight planets, it formed many more, and collisions between those planets under construction were relatively common occurrences. A particularly bad collision or series of collisions could have caused such an energetic punch to the growing Venus or Uranus that the planets' entire orientation was shifted, leaving them out of sync with the rest of their spinning siblings.

## Collisions between those planets under construction were relatively common occurrences.

### HOW EARTH SPINS

You can remember which way Earth is rotating by thinking about the time zones: the farther east you go, the later it is—they're pushed toward daylight sooner than the west.

### VENUS
243 days,
26 minutes

*Venus spins
clockwise. As far
as we can tell, this
means Venus is
upside down.*

### MERCURY
58 days, 15 hours,
30 minutes

### JUPITER
9 hours,
55 minutes

### EARTH
24 hours

### MARS
24 hours,
40 minutes

### SATURN
10 hours,
36 minutes

### URANUS
17 hours, 14 minutes

*Uranus rotates 90 degrees
off from everything else,
like a bead rolled along
the ground, instead
of spinning vertically.*

### NEPTUNE
16 hours,
6 minutes

## HOW THE PLANETS SPIN
If you consider the plane of all the planets'
orbits around the Sun as a flat surface, most
planets spin as though they were a coin spun
on its edge, flicked counterclockwise.

▲ **Rotation of planets in the solar system**
Each planet's direction of rotation, and
the time taken to make one complete
rotation. In this schematic, we have a
bird's-eye view of the solar system, though
the more familiar view of the planets from
the side is shown.

# HOW DID THE MAJOR PLANETS BUILD UP THEIR MASS?

Fortunately for life on Earth, the kinds of collisions that might have tilted Uranus and Venus are a thing of the distant past. As material from the disk of gas and dust collected into small objects that might eventually become planets, the early solar system was cooling down. Unlike a gas cloud, which can retain heat a little better, individual balls of rock are not so good at keeping the entire solar system warm.

At some point, the solar system had cooled to such a point that it became difficult for the newly formed rocks to stay hot. That heat had allowed these rocks to stick together more easily, instead of bouncing or shattering, as the rocky material was pliable, like molten glass. Without this stickiness, the growth of most objects stopped. To keep growing from this point, the larger objects had to collide with others—but many of these smaller pieces have remained just as they were at the end of this phase. Asteroids and comets are among these remnants of the formation of the solar system, and date back some 4.5 billion years.

Billions of years have passed since those earliest days, and most of the large objects in our solar system have either already had the collisions they're going to have, destroying themselves in the process, or they've survived such encounters and have found a stable place to orbit the Sun, free of disturbance. This disturbance-free orbiting is such a feature of the major planets in our solar system that it has now been folded into the definition of a major planet—no other similarly sized objects should cross the orbit of a major planet.

### PARTIAL "PLANETS"

There are a few objects that seem to have done okay in building up mass to the point of making themselves round, and also managed to avoid being destroyed by colliding with a larger planet, but then failed to continue gaining mass to the point where there weren't any other objects in their orbit. These are what we now call dwarf planets, or minor planets. They made it partway to being a planet, but they didn't quite get all the way there. (See the sections about Pluto on pages 78–81.) The pieces of our earliest solar system that never quite made it into a full-sized planet, the asteroids, can be scientifically fascinating. These objects serve as time capsules to the formation of the early solar system, telling us what it was made of and how fast it was cooling down, among other things.

In order to get a smaller object to stick to a young, growing planet and gain its mass, the force of the collision is the deciding factor. This is controlled by two things—the speed at which the two objects

> This disturbance-free orbiting is such a feature of the major planets in our solar system that it has now been folded into the definition of a major planet.

▲ An illustration of the rock and metallic asteroid Psyche. It is currently too distant to photograph, but in 2026 we will be able to take real images of this small world.

collide, and the mass of the two objects. If the two objects are the same mass, but collide at very high speed, they both may shatter. This kind of head-on collision is what we think happened to create an X-shaped cloud of debris around an asteroid observed by Hubble in 2010. You can just as easily shatter a smaller object by running it into a much larger object. This is the same as breaking a rock by dropping it on the ground from a height.

The shards of shattered objects from the early solar system can end up wandering the solar system, and many of them are found in the main asteroid belt between Mars and Jupiter. There's another reservoir of asteroids and comets that sits out beyond the planet Neptune.

## METEORITES
We encounter the smallest pieces of the early solar system with every meteor shower, and very occasionally, one larger piece will survive its passage down through the atmosphere and land somewhere on Earth. From the analysis of these pieces, we've learned quite a lot about the early solar system, and what metals were present there. See pages 60–61 for more on meteorites.

# WHAT ARE METEORITES MADE OF?

The meteorites found on Earth are usually made of some combination of rock and/or iron. With the infinite inventiveness of astronomers, we duly classified them as "stony," "iron," or "stony-iron," if they're a relatively even blend of the two. Stony meteorites can (and often do) contain some small fraction of iron, but only those meteorites that are almost entirely metal are classified as iron meteorites.

The vast majority—over 90 per cent—of meteorites that fall to Earth are stony. The problem with spotting stony meteorites is that they look like rocks, and there are a lot of rocks on our parent planet. The only place where it's easier to spot this type of meteorite is where you wouldn't expect to find many rocks naturally. The best place to look for stony meteorites is therefore the glaciers of Antarctica. Glaciers do not normally have rocks on top of them, and the ice there doesn't move very rapidly, so any meteorites that fall there will both stand out against a bright white background and stay put for quite a long time. These meteorites are some of the oldest untouched rocks in the solar system, and so are profoundly interesting to scientists as conveniently delivered time capsules from the very beginnings of that system.

There's one other interesting set of stony meteorites—we have assembled a collection of about 30 meteorites that have come to Earth from Mars. These are chunks of the surface of Mars that were flung away after a larger object hit the red planet, and then wandered the solar system until they encountered Earth. These Martian meteorites tend to be much younger than the rest and are composed of a different set of elements and minerals, so we can tell them apart from the older stony meteorites without too much trouble.

▲ A metallic meteorite (top) is notably brighter and shinier compared to its stony counterpart (center), due to the high levels of iron and nickel it contains. The stony counterpart has scorch marks that are a record of its journey through the atmosphere. Bottom is a type of meteorite called a pallasite; a stony-iron meteorite, it is largely iron, but has inclusions of olivine (also known as peridot).

Metal meteorites are less affected by erosion, so they can be identified for a much longer period of time after they fall.

### METAL-DETECTING

Iron meteorites are much easier to spot than the stony ones. This is partly because it's relatively unusual to find weathered chunks of iron sitting on the ground, and so it's easier to recognize that they are out of place. As with the stony meteorites, they are most easily found in deserts and in the glaciers of Antarctica, where they are likely to stand out more from their background, but they can also be found in other places. These metal meteorites are also less affected by erosion, so they can be identified for a much longer period of time after they fall.

The metal in these meteorites is not entirely iron, though that's a large fraction of it—the remainder is nickel. This blend of metals is part of what makes them unique. Another part is that they tend to have long crystals of metal within them; with a slice and polish of the meteorite, we can show these crystals off. (Without this extra treatment, they look much less dramatic.)

Iron meteorites and stony meteorites also behave somewhat differently as they come through the atmosphere. The stony ones tend to be more fragile against the force of the atmosphere, and as a result they can shatter more easily before they hit the ground. Iron meteorites are more resilient. The heat of reentry doesn't usually stress the metal to its fracture point, but it can certainly reach that point for stone.

### OPPORTUNITY ROCKS

In 2005, the Mars rover Opportunity found an iron meteorite on Mars—the first time we'd found a meteorite on a planet other than our own. Its irregular, pitted, and lumpy look is much more typical of the appearance of a nickel-iron meteorite. Opportunity found several more of these meteorites, including the example above from 2010.

The Chelyabinsk meteor that exploded over Russia in 2013 (see page 18) also left behind some pieces for us to examine; the largest of these were extracted from a lake. Because it exploded so dramatically in mid-air, it had been suspected to be a stony type object, and the pieces confirmed that hunch.

Scientists like collecting as many meteorites as they can, because differences between them help us to understand how uniform the disk of gas and dust might have been when the solar system began to form, or if there were interesting variations in how much of which elements were present. This kind of information helps us refine our understanding of how planets form.

# IS GOLD REALLY MADE IN STARS?

Asteroids and planets, unlike the stars, have no fusion occurring at their cores (see pages 98–101) and thus cannot make any new elements. This means that all the metals present on Earth and in all the asteroids were already there, in the gas cloud that our star formed out of, before the planets began to form. This is interesting, because by all current models our universe began with a pretty limited store of elements.

Hydrogen, a little helium, and a tiny bit of lithium were the only elements present at the beginning. They're the three lightest elements in the periodic table, and so every element heavier than hydrogen and helium has been added later into the universal mixture—and they've been added by stars. We'll go into more detail on how the stars can build up so many more elements in the next chapter, but the key point for now is that without the presence of stars in the area before our solar system had formed, all the carbon, silver, and gold we find on our planet would not exist.

The iron and nickel that make up the core of planet Earth and the metallic meteorites are both formed in the very centers of large stars, at the very end of their lifetimes. All of the stars that are capable of creating nickel and iron go through a supernova explosion at the end of their lives, triggered when nothing remains in the core of the star that can be burned. (More on this in Chapter 4.) The burn that creates nickel is the very last one possible, so a supernova is imminent once this process starts.

> Without the presence of stars in the area before our solar system had formed, all the carbon, silver, and gold we find on our planet would not exist.

Supernovae are pretty sizable explosions and spread some of the elements they've created into the space that surrounds them. Over many generations of stars, the small amounts of more complex elements that have been created and dispersed build up, to the point that any given gas cloud has probably been filtered through a few stars at some point in the past. So it was with the cloud of gas that became our solar system—the gas had already been recycled through a few rounds of stars, which had filled the gas with a good amount of nickel, iron, and all the other elements we find on our planet—including gold.

◄ A pure gold nugget, from Kalgoorlie-Boulder in Australia. Few metals are found in their pure elemental form on Earth; gold is one of a select few.

▲ As seen by the Hubble Space Telescope,
this image shows a colorful portion of the
Veil Nebula, the remains of a supernova
that may have exploded some 10,000
years ago.

# Special delivery: contains living beings

If you wanted to naturally transport life around a solar system, how could you do it? We know that pieces of the early solar system are regularly smashing into Earth, and we know that they've also smashed into Earth's sibling planets. It's been suggested, therefore, that the meteorites which make it to the surface of a planet could carry life, or the materials for life, buried deep within them—a theory dubbed "panspermia." Earth has been on the receiving end of some Martian rocks, so perhaps once one planet develops complex organic materials, it can share these with its siblings via meteorites. But how likely is that? And could life survive this kind of transport?

Perhaps, once life arises somewhere in the universe, an impact upon that planet could create such an explosion that pieces of rock carrying bacteria are flung into space, where they cruise about until crash-landing on another life-friendly planet. To work out whether this theory has legs, we'll need ponder how rocks might be thrown out into space, and then look in more detail at what might happen to bacteria on an interplanetary road trip.

▼ **Extreme living**
A colorized view of the archaea *Methanococcoides burtonii*. These cold-loving archaea were discovered in 1992 in Ace Lake, Antarctica, and can survive in temperatures as low as 27.5°F (–2.5°C). Extremophiles such as these may be the most hardy forms of life for any kind of interplanetary transport.

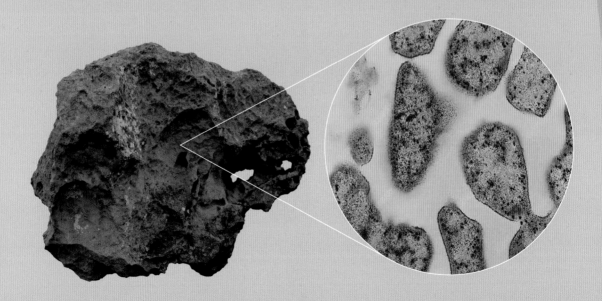

### 1. How can you launch the messenger rock?

It's an inelegant solution, but the best way to launch a small chunk of the surface of some planet out into space is to punch that planet extremely hard with another piece of rock. An asteroid is probably the best bet, as we have lots of them in the solar system, and they're large enough to create a high energy impact crater—which in turn may give enough energy to some small chunks of rock to fling them entirely away from the planet. The arrival mechanism is then the exact same; wait for this rock to smash into the surface of another planet.

### 2. Would life survive the journey?

We still don't have a complete answer to whether or not bacterial life is hardy enough to survive this entire process, particularly when we include the blasting off the original planet and the landing on the new planet stages of this process. If it can, then this might be one way to get bacterial life from a planet to its siblings within a solar system, however unlikely it might be for those pieces of rock to be flung in exactly the right directions. However, scientists have been working on figuring out the plausibility of the middle bit (sailing through the vacuum of space) by sticking microbes or microbe-bearing rocks on the outside of the ISS, and seeing whether they survive.

A few microbes from the town of Beer in southwest England managed to partially survive a 533-day stint on the outside of the International Space Station. This is impressive, but you should take note that it was a partial survival of one strain of bacteria, on a relatively short exposure time for a between planets unplanned journey. There's no guarantee that a rock blasted off the surface of another planet will be covered with a particularly space-hardy breed of microbe, nor that the meteoroids will arrive at another planet in a short period of time.

It might instead be easier to transport not the live bacteria, but the organic molecules that are required to build them. Amino acids are one such set of complex organic molecules, and are often called the building blocks of life. If a comet or meteorite can bring amino acids to a planet, it might help to jump-start the development of life on that planet by skipping the steps required to build those molecules from scratch. These molecules don't need to have formed on the surface of another life-bearing planet—surprisingly complex molecules can form in the gas between stars in our galaxy. Because these simpler building blocks may be more widely available than planets with life that are also being bombarded with impacts, it may be easier to distribute these molecules than to distribute entire bacteria. If you're looking to use meteorites or comets as your delivery mechanism of the building blocks, you'd also only have to wait for one impact, rather than two.

### Verdict

Until we have more concrete proof of either of these processes happening out there, they will remain intriguing possibilities, but we're working toward testing each of the steps individually to get a sense of how plausible all of the steps together might be. In general, the more impacts you have to wait for, the longer (and less likely) the process becomes, so it may turn out that simply waiting for comets to smash into your planet bearing complex molecules on ice is the easiest way to speed up the process of building up the ingredients for life.

# IS THERE LIFE ON MARS?

We humans have sent a lot of robotic explorers to Mars, and with a number of rovers on its surface we have an incredibly detailed view of what the red planet looks like up close. The rovers have gathered a great body of evidence to tell us that there was very likely a lot of warm, life-friendly water on the surface of Mars for quite some time.

The Opportunity rover in 2011 found a huge streak of gypsum sticking out of the surface, along with haematite "blueberries"—both signs of a wetter past on Mars. There's nothing as far as we can tell that might have entirely prevented life from arising on Mars 1 billion years ago, but equally we haven't proved definitively that there was life, either.

However, the current environment on Mars is very different from its watery past—it's now a cold and cancer-inducing place for humans. There's still water there, but it is usually in the form of ice. The polar ice caps on Mars are both substantial in size and mostly water ice. These grow and shrink with the seasons, just like our own ice caps. Mars has

such dramatic dust storms that the entire planet can be engulfed in them, hiding the features of the surface from our view.

As wonderful as the images being returned from the rovers exploring Mars are, they present a very biased view of what the surface looks like. The rovers have been sent to rather flat ground, which gives relatively familiar-looking horizons, instead of something jagged and foreign. There's an important reason for this: we can't land rovers on very rocky ground or at high elevation. We need landing sites to be at low elevation so that there's a lot of atmosphere to help slow the rover down, and we want the ground to be flat so we don't accidentally drop a very expensive machine on a boulder. These restrictions on our landers mean that we miss out on the most dramatic vistas Mars has to offer.

Valles Marineris, shown on the right, is one of these places, and it makes the Grand Canyon in the United States look unimpressive. At more than four times as deep, and five times as long, Valles Marineris would be challenging territory even for humans. Trying to land something the size of a large car on the jagged ground there is impossible for now. Water very likely flowed through it, but it's unclear whether it was formed by the water itself, or whether it is a more fundamental crack in the planet's crust. In any case, it is the solar system's largest canyon. What's more, Mars also has the solar system's biggest volcano: Olympus Mons (shown on the next page), one of a series of extinct Martian volcanoes. At 14 miles (22.5 kilometers) high, it's three times higher

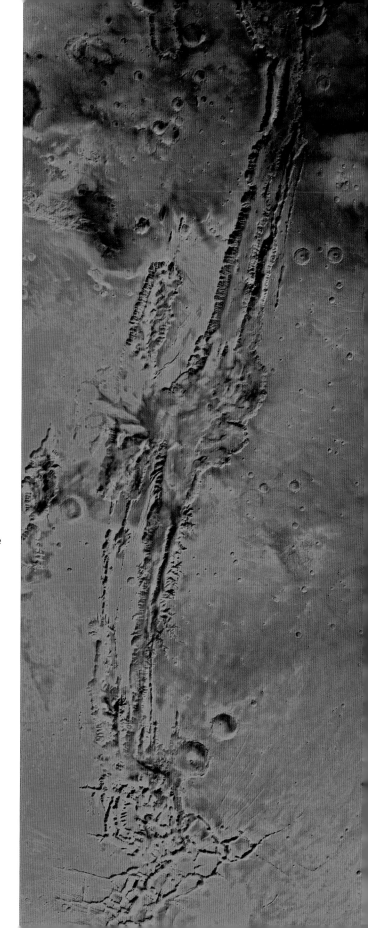

▶ Valles Marineris, the great canyon of Mars. The scene shows the entire canyon system, over 1,800 miles (3,000 kilometers) long and averaging 5 miles (8 kilometers) deep.

◀ A "selfie" taken by the Mars 2020 rover, Perseverance, which is hunting for signs of ancient microbial life in Jezero crater on Mars.

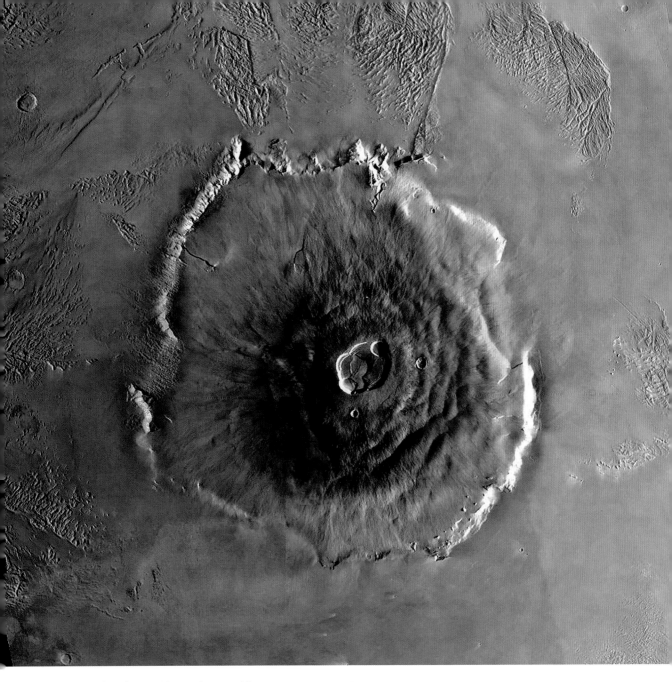

▲ The Olympus Mons volcano on Mars, as seen by the *Viking 1* orbiter. Olympus Mons is around 370 miles (600 kilometers) in diameter and the summit caldera is around 15 miles (24 kilometers) above the surrounding plains. This is so high that it is elevated above a substantial fraction of the Martian atmosphere, making a parachute-slowed descent for any rover impossible.

than Mount Everest, and twice as big as Mauna Loa in Hawai'i, if you start measuring from the ocean floor. Olympus Mons has cliffs several miles high right at the base, which will make for an incredible climb whenever we get people over to Mars, but is a rather severe impediment to getting rovers onto the volcano. (This particular problem seems to be unique to Olympus Mons—the smaller volcanoes don't have the same kind of cliffs.)

## VEINS OF GOLD

The presence of volcanoes on the surface of Mars brings up another intriguing possibility. On Earth, volcanoes serve to dredge up some of the heavier metals that would have sunk to the center of the planet while everything was still molten, and before the crust had solidified. On Earth, these upwellings are responsible for lifting precious metals such as gold back up to the surface, as it is dissolved in with the rock and volcanic lava. We can also get some additional gold at the surface from impacts from asteroids, but at least on Earth, it seems that volcanic regions are better places to find gold.

As a sibling to Earth, and having formed from the same mixture of materials, Mars should also have roughly the same fraction of gold as Earth does, and it has the volcanoes that could bring the gold back up to the surface. While Earth's volcanoes are active now, Mars' were active 1 billion years ago, but otherwise it seems reasonable to think that this redepositing of gold on or near the surface may well have also happened on Mars.

Volcanoes don't just bring gold up to Earth's surface—they also produce geodes. Geodes are hollowed-out bits of rock, with crystal formations growing on the inside. The type of lava that flows smoothly (instead of exploding) frequently forms small bubbles. Sometimes these bubbles pop, but sometimes they harden in place. If the bubble survives, over time, water may be able to seep inside, depositing minerals, which can form crystals. We know now that Mars had plenty of water at one point, so this process may also have been possible on Mars.

Agate also has a volcanic origin. Agate forms when material full of silica (the main ingredient in white sand and glass) slowly fills or partially fills an irregular hole, perhaps formed by a gas bubble in lava. As the material fills in the hole, it creates the bands you see in some agate, like a tree produces bands as it grows.

The major stumbling block in suggesting that there might be geodes, agate and gold on the surface of Mars is simply that we'd need to go near the volcanoes to find them, and as we've said, that's a particularly dangerous place to try to put a rover. It has proven difficult enough to get any kind of machine to the surface of Mars, so making a landing extra precarious has not been at the top of anyone's priority list. Until we can figure out a safe way to explore the volcanic regions of Mars, either with rovers or a human crew, the presence of gold, agates, and geodes will remain simply plausible—and the prospect of Martian gold jewelry a far-off possibility.

## PLATE TECTONICS

Segments of Earth's crust drift relative to each other, colliding and spreading apart at the various boundaries between segments. This is known as plate tectonics. Where individual segments (plates) collide, mountains and often volcanoes are built. Where they separate, deep rifts can arise. Planets without plate tectonics must build mountains, valleys, and volcanoes through other means.

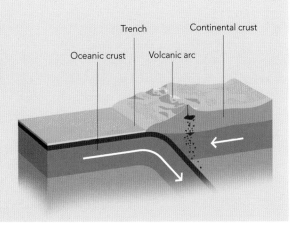

Trench

Continental crust

Oceanic crust

Volcanic arc

# HOW DO WE KEEP FROM CONTAMINATING OTHER PLANETS?

Given all the hints of a wetter, possibly habitable past on Mars, it makes sense as we continue to explore the planet to avoid contaminating its surface with our rovers. And indeed, contamination of other worlds is a major source of concern, particularly when we're going someplace that might be habitable for some form of life.

Every spacecraft (including satellites, telescopes, and rovers) has all of its pieces go through some pretty serious decontamination before launch, and most are assembled in a very high-test clean room. (The last time I was near a clean room of this caliber, we weren't even allowed in the corridor that led to the clean room, in order to try to keep airborne particles to a minimum.)

Part of this clean assembly is for the good of the craft itself. Discovering that, somehow, dust had settled all over the mirror of your new telescope would be a nightmare scenario, meaning that it's seriously under-performing, to the detriment of all the science that it might have been able to do. And you certainly don't want grit making its way into the moving parts, which could jam their operation. If the part that needs to move is critical to the success of the mission (say, for instance, it's your communications antenna), a stuck part could derail the entire project.

If your spacecraft has a gas chromatograph or a mass spectrometer on it, both of which are instruments designed to sample the molecular properties of a gas, then you have to be extra careful. If you have a sensitive tool measuring the chemical composition of the air or vaporized soil, any residuals from Earth on the tray you're using to hold the gas or dust will contaminate your measurement, and you'll wind up measuring the contaminant instead of what you really want it to be sampling (i.e., Mars). This has actually been an issue in the past: the *Viking* landers on Mars in

1976 made this kind of measurement of the soil and found some unusual chemicals in it. Unfortunately, it was found that this reading taken from the soil measurement could be produced either by some interesting chemistry on Mars itself, or by some residual cleanser on the soil tray.

If we add in some later measurements by the Phoenix lander in 2008, it's possible that the "interesting chemistry" interpretation is the correct one, but until some more measurements are made on the exact isotopes (see the box, opposite) of the atoms involved in the reading, it will be difficult to rule out the cleanser contamination explanation.

The Curiosity rover, on the other hand, seems to have found genuine complex molecules from Mars' air and vaporized rock, but the scientists working on the team could only say this after determining that the results of the experiment on different rocks were sufficiently dissimilar to rule out contamination. If you thought the signal you were getting was from a single contaminant on your tray, it would produce a similar signal no matter which rocks you were inspecting, so finding different signals from different rocks means that it's probably something attached to the rocks, and not to your measuring device.

Landers so far have been placed on parts of Mars and in other places in the solar system that are pretty dry and reasonably flat. The dry tends to go along with the flat; by and large our best observations of liquid material flowing on Mars have been along steep slopes, where it is too

▲ The Curiosity rover, seen on its 2,553rd Martian day, took this image of itself as a mosaic of 57 separate images.

dangerous to land. Avoiding those places for the purposes of avoiding contaminating any water reservoirs is done out of an abundance of caution. It's hard to guarantee that we don't have any cleanser products or other contaminants left on our rovers, in spite of our best efforts to send as clean a craft as possible. If there is some form of extreme life on Mars, we'd like to not kill it accidentally.

## ISOTOPES

Not all copies of an element have the same number of neutrons (see page 100) in their cores. These differently constructed atoms are called isotopes of each other. They remain the same element, because they contain the same number of protons in the core of the atom, but the mass of the atom changes.

# WHAT CAUSES JUPITER'S STRIPES?

Jupiter is the largest planet in our solar system—a tremendous assembly of gas, now thought to have a diffuse collection of heavier elements in its core. One model suggests what might have once been a rocky core was destroyed early in Jupiter's lifetime by a massive collision. Of all the outer planets, Jupiter's cloud layers are the most distinctive, with dramatic currents and eddies of different colored gasses.

Jupiter is classed as a gas giant, as are the other planets in the outer solar system, and it is more massive than the rest of the planets in the solar system combined. It's also host to an enormous, hundreds-of-years-long storm, which we know as the Great Red Spot. And crossing the planet, from top to bottom, are cloud bands, alternating in color (shown opposite).

Jupiter's atmosphere is doing some pretty weird things, and we don't fully understand all of what's going on. We know that the light stripes and dark stripes are made of slightly different gases. At the boundaries of these stripes are narrow jets of high wind, which push the nearby atmosphere around with it; the bands themselves are relatively stable.

The light stripes seem to be made of cold gas that is coming up toward the surface of the atmosphere, and the dark stripes are warmer gas, sinking down toward the center of the planet. The light ones are light because there's a lot of ammonia in the upper atmosphere of Jupiter, and as it cools, it forms pale clouds, like the clouds in our own sky. If the gas warms up, the clouds will disappear, and what we're seeing as dark bands are actually a deeper, darker layer of clouds. Jupiter's still doing something unusual with those bands, because the Sun also has gas rising and falling with different temperatures, but it doesn't have any such stripes.

## WIND ASSISTED

At a very basic level, Jupiter's stripes are present because there are jets of wind running around the planet. The jets form a boundary for the gas, and the gas is easily redirected along the path of the wind. The jets alternate in direction as you go from the equator to the poles of Jupiter, which means the atmosphere is being pushed in different directions, depending on where you are on the planet. Eddies form at the edges, the way swirls are produced if you push your hand backward against the flow of water. The wind jets mark the edges of the bands, and each stripe moves in a different direction. The part that we don't yet understand is why those jets exist in the first place.

Broadly, there are two main ideas. One is that this is turbulence at the surface level, like clouds in the upper atmosphere of Earth. Perhaps there was some turbulence—a little bit of bumpy air—and it ran into another patch. As patches of turbulence catch up to each other, they can combine in what's called a cascade. If there's a constant source of the little eddies of turbulence, then you can maintain bigger turbulence (like the wind jets) just by tossing the little ones together. But while this method can create the jets, it's not great at keeping them stable.

▶ Jupiter's southern hemisphere, and the Great Red Spot, photographed by NASA's *Juno* spacecraft as it performed a close pass of the gas giant planet. Swirls of cloud layers of different compositions (and therefore colors), are emphasized in this color-enhanced image.

Since Jupiter is rotating relatively quickly (a Jovian day is just under 10 hours long), and seems to be all gas, the gas could alternatively form cylindrical shells of material that rotate in different directions, as you go out from the center of the rotation. Planets are not cylinders, but if you start with a cylinder, and carve into it at the top and bottom, you can create a sphere. So these rotating bands might appear as the surface of Jupiter cuts into different layers, as you get closer to the poles. This is similar to the way that a sharpened wood pencil has a stripe of pencil lead, a stripe of the wood that's been cut into and the outer layer of the pencil that wasn't touched, but where each layer of the pencil is rotating. This is also not a perfect solution, because generally it doesn't form enough bands to match the Jupiter we see.

▲ Jovian clouds are seen here in striking shades of blue. The *Juno* spacecraft captured this image when the spacecraft was only 11,747 miles (18,906 kilometers) from the tops of Jupiter's clouds. Differences in the heights of the cloud layers cast shadows on the lower layers, creating a dramatic textured effect.

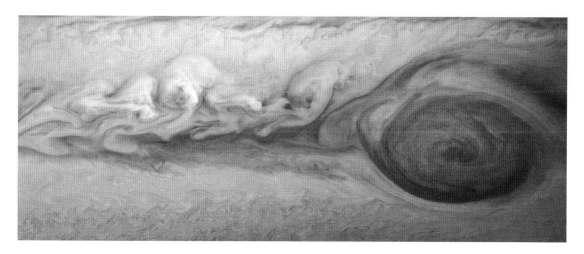

Since neither of the explanations for the stripes comprehensively resolves the uncertainty around Jupiter's atmosphere, for the moment we have to stick with observing the banded clouds and the jets that drive them, in the hope that understanding the details of their behavior will hint toward one answer over another. The *Juno* probe, which arrived in orbit in July 2016, is helping to answer these questions as it orbits Jupiter.

## GREAT RED MYSTERY

Complicating matters is the Great Red Spot, nestled in between these bands of high winds. The Great Red Spot is more like a hurricane than anything else, but with reliable observations of this storm spanning hundreds of years (since the 1800s), and at several times the size of our entire planet, it's both the longest-lived and the most physically expansive storm we know of. It's also a very perplexing storm, and explanations of its longevity and its size are hampered by how different Jupiter is from our Earth. A planet with no solid surfaces under its clouds may well be able to maintain hurricanelike storms for so long, but even the color of the Great Red Spot is not very well understood. Something must be happening with the gases that make up the storm to produce such a vivid color, but exactly what that is remains a bit of a mystery.

▲ This *Voyager 2* image shows the Great Red Spot, with its fine structure clearly visible. This storm, larger than Earth, seems to be able to heat the atmosphere above it by about 700°F (370°C).

▼ The *Juno* spacecraft was launched in 2011 with the aim of enhancing our understanding of Jupiter's structure.

# To see a sky with more stars than Earth's, should you try Jupiter or Deimos?

In order to assess where the greatest number of stars would be seen, we'll need to work out how much of the sky is visible in each case. Given that Jupiter is 1,300 times bigger than Earth, and larger still than the Martian moon Deimos, we should also consider what role size plays.

There are a few pieces to this puzzle. The first is to understand why we see the number of stars that we do here on Earth. The second is to examine how this might change if we placed ourselves on a world 1,300 times larger than our rocky Earth. Finally, we'll look at a significantly smaller alternative: one of Mars' moons, Deimos. We'll ignore the differences in rotation speed, which controls how fast the stars rise and set, and focus only on how many stars you can see in a given moment.

## 1. Stargazing from Earth

On Earth, we see a fixed number of stars at night because there is, at best, only 180 degrees of sky to look at. Of a full circle, half falls below our feet, and half is sky. The ground of our planet appears flat to us because the curve of Earth's sphere is so gentle. Our half-circle of sky is what falls above our local, mostly flat, surface. As Earth revolves, our personal patch of surface points toward different stars in the sky, which causes some stars to rise and others to set in our visible sky.

## THE TINY IRREGULAR MOONS OF MARS

Deimos and Mars' other moon, Phobos, are interesting objects; they're both irregularly shaped, and seem to have a lot in common with asteroids, leading some planetary scientists to suggest that they might be captured asteroids. Our own Moon was formed after a roughly Mars-sized object struck the proto-Earth (see page 34), early in the formation of the solar system. While it's possible that Phobos and Deimos formed this way, their smaller size means that it would have been a less catastrophic collision. Phobos is about twice as large as Deimos, and is much more intensely studied; it orbits Mars about three times for every Martian day, and is slowly falling toward the surface of the red planet. It's tidally locked to Mars, as our Moon is to Earth. If you wanted to have a very rapidly changing sky, without moving at all, you could sit yourself on the far side of Phobos, pointed away from Mars, and experience a sunrise every 7 hours.

## 2. Moving north or south on Earth

As you move north or south on the planet, the position of the stars shifts in the sky, because you're pointed at a different patch of stars. Most notably (in the northern hemisphere), the North Star will appear higher or lower in the sky as you move north or south respectively. If you moved from the North Pole to the South Pole, the stars you would see would be entirely different, as you'd have no overlap between the 180 degrees of sky you're pointed toward at the North Pole, and the 180 degrees of sky at the South Pole.

## 3. Stargazing from Jupiter

Going from Earth to the much larger Jupiter, the only change is that the curve of Jupiter's sphere is even gentler than Earth's. The atmosphere below you would still locally look flat, meaning that you'd still have 180 degrees of sky and 180 degrees of planet under your feet. However, because the sphere of the planet is so much larger, it would require a longer distance to change your angle on the stars. So, you would have to travel much farther to notice that the North Star had moved.

## 4. View from a small planet

The only way you'd be able to see more stars is if you were on such a small planet or moon that you could see the curve of the planet dropping away from you. Here you would have more than 180 degrees of sky, because your horizon would always be a circle below you. You would be able to see stars up, horizontally, and at an angle down beyond your feet. Deimos, one of Mars' small moons, which is only about 8 miles (12 kilometers) across, might work.

## Verdict

Jupiter won't give you a more expansive sky, in spite of its tremendous size. In fact, the only change Jupiter's size would make on the sky you see is that it would take much more effort to change your view.

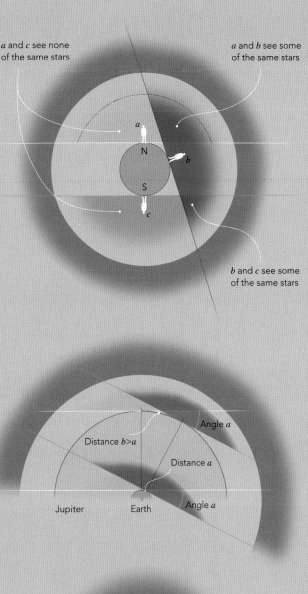

*a* and *c* see none of the same stars

*a* and *b* see some of the same stars

*b* and *c* see some of the same stars

Angle *a*

Distance *b>a*

Distance *a*

Jupiter    Earth    Angle *a*

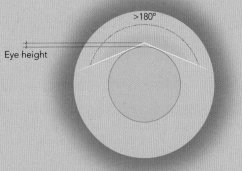

>180°

Eye height

# WHY IS PLUTO NO LONGER A MAJOR PLANET?

In 2006, the International Astronomical Union (IAU) made a change to the set of criteria used to categorize major planets, and Pluto was reclassified as a dwarf planet. For a tiny object at the edges of the solar system, the decision to reclassify Pluto ruffled more feathers than one might have expected. Thanks to *New Horizons*' flyby in 2015 we know more about Pluto than ever before, and it is undeniably a fascinating little world.

The truth is, Pluto never fitted in very well with the other eight planets. The orbits of all the major planets around the Sun lie very flat, compressed into synchronicity. The orbit of Pluto in comparison is wildly askew; it's tilted by 17 degrees relative to the rest of them. Also, Pluto's orbit is extremely elongated, unlike the other planets, which travel around the Sun in almost perfect circles. What's more, Pluto is quite small. It's big enough at 1,475 miles (2,374 kilometers) across to have compressed itself into being a sphere (compared to the lighter and more irregularly-shaped asteroids), but it has a moon, Charon, that is almost as big as itself. None of the inner eight planets behave this way.

While we had noted these irregularities with Pluto's behavior, for a while it seemed to be a unique little world, surrounded by much smaller objects than itself. It seemed sensible to leave Pluto as a planet, as long as it remained a one-off. This is where we started running into problems. Our ability to detect Pluto-sized objects got a lot better, and all of a sudden we had found a handful of other worlds. Most notably, there was Eris, which is even farther out from the Sun than Pluto, and also bigger.

Now a decision had to be made: either we allowed Eris into the solar system as the tenth planet (if Pluto was a planet, Eris surely was), or we would have to come up with a clearer definition of what a planet is. The problem with adding Eris and keeping Pluto as planets was primarily that Eris' discovery was proof that Pluto was not unique; it was merely the first in a class of objects we'd previously been unable to detect. That meant that we'd be constantly adding new planets to the list of siblings that make up our planetary family as

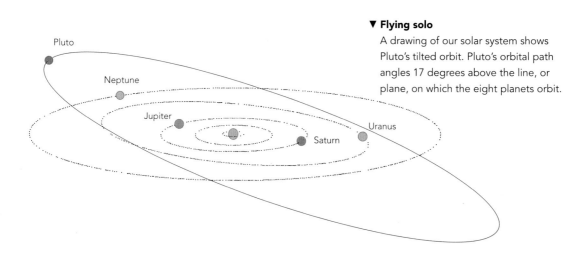

**▼ Flying solo**

A drawing of our solar system shows Pluto's tilted orbit. Pluto's orbital path angles 17 degrees above the line, or plane, on which the eight planets orbit.

more and more of them were discovered. In response to this problem, the IAU decided to impose a set of three criteria that any object in the solar system must meet in order to qualify as a planet (see the box on the right). Pluto and Eris failed the last of these criteria, and they were subsequently redesignated as dwarf (or minor) planets. This classification distinguished them from the irregularly shaped objects that orbit the Sun (like the asteroids), but swept them out of the main planet category.

The decision to reclassify Pluto was a conscious choice to make our definitions more consistent. Our understanding of the solar system will gradually become more complex as time goes on, so it makes sense to let our classifications reflect the complexity of our planetary family.

## WHAT MAKES A PLANET?

At the IAU 2006 General Assembly more than 2500 astronomers from around the globe agreed to a new system for classifying planets:

- A planet must orbit the Sun, and not another, smaller object. (It can't be a moon.)
- It must also be sufficiently massive to have compressed itself into a sphere, which excludes all the asteroids and also the funny-shaped comets.
- Finally, it should have cleared the area around its orbit of other objects. This means that there shouldn't be other worlds of a similar size anywhere along its orbit.

▼ Pluto's surface sports a remarkable range of subtle colors, enhanced in this view to a range of pale blues, yellows, oranges, and deep reds.

# WHAT DO WE KNOW ABOUT PLUTO?

While Pluto was being reassigned to the dwarf planet side of the family, very little was known about what it actually looked like. This all changed when the *New Horizons* spacecraft made it to Pluto, after traveling for a decade to get there.

Before *New Horizons*, we could tell you that Pluto seemed to be made of a roughly even mixture of rock and ice, and that it had an abundance of moons. In addition to its large companion, Charon, which is 10 per cent of Pluto's mass (this is huge for a moon), Pluto has four other moons: Nix, Hydra, Kerberos, and Styx.

The only images of Pluto to be had were from a combination of Earth-based observatories and some space-based, like Hubble. Pluto was still too far away and too small to do much with—the highest resolution images that Hubble could return were still very fuzzy. The resolution was so poor that we could really only tell that there were variations in brightness and color, instead of it being more consistent all over, like many of the icy moons of Jupiter or Saturn.

Sending *New Horizons* was a long-term project, and a risky one. On top of the 10-year journey to get there, it was unknown if there would be hazardous material surrounding Pluto—another, smaller moon, or a dense debris field, perhaps. But *New Horizons* succeeded brilliantly, and in doing so, truly revolutionized the way we see Pluto. The images that returned were the first high-resolution, color images of the world at the edge of our solar system, providing puzzles, answers, and more. Pluto went from a fuzzy blob to a fully developed little world nearly overnight (see the image opposite).

While we had known that Pluto had some color variation, no one could have predicted the terrain that *New Horizons* showed us. There was a bright, heart-shaped plain, oddly missing any sign of cratering, the dark regions immediately next to it and the valleys and mountains.

## GEOLOGICAL ACTIVITY?

A lack of craters means that Pluto's surface must be young. This surface has somehow erased all memory of impacts with other objects, which have almost certainly happened, meaning that Pluto is geologically active. Given its distance from the Sun, there shouldn't be enough heat to warm the surface of the dwarf planet. It's possible Pluto could have retained some heat from a high-energy collision, but we would have thought that those impacts would have happened a long time ago, and therefore the heat should have already vanished into outer space. Pluto being the way it is means that one of these two assumptions is wrong, or there's another, third option we haven't thought up yet.

In addition to the oddly flat plains, Pluto has mountains, some of which are higher than 11,000 feet (around 3,350 meters). Given how high and steep they are, and the materials present on Pluto's surface, they must be made of water ice. No other type of ice would remain so mountainous—it would have slumped with time. There's a variety of ices to work with: for instance, nitrogen, which is rarely an ice on Earth (even in low-temperature chemistry experiments it's used as a coolant in its liquid form), is present as an ice all over the surface of Pluto. It seems that the nitrogen ices act on Pluto as glaciers do on Earth, flowing slowly along the surface. Pluto showed us an incredibly diverse and unexpectedly alive surface via *New Horizons*.

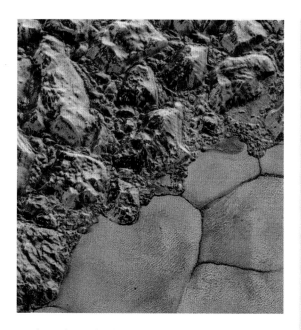

## A LAYERED ATMOSPHERE

Pluto's atmosphere was another surprise—very little was known about it, as trying to observe it from Earth is extremely hard. The atmosphere we observed with *New Horizons* was not only much bigger than we thought (it extends much farther from the surface than expected), but it has bizarre stripes in it. That's not an artifact of the image—those are truly in the atmosphere. It seems that some of the surface ice evaporates during daytime, and suspends itself in the atmosphere, where it lingers for a long enough time to build up these layers. If any gas escaped immediately to outer space, as we had thought it should, then Pluto's atmosphere should be tiny, a thin shell of gas sheathing a mountainous, glacier-y, young world. But we now know differently.

▲ This enhanced color mosaic shows the al-Idrisi mountains meeting the shoreline of Pluto's "heart" feature. Although a small snapshot, it shows us Pluto's extremely varied terrain.

▼ Just 15 minutes after its closest approach to Pluto on July 14, 2015, NASA's *New Horizons* spacecraft looked back toward the Sun and captured a near-sunset view of the rugged, icy mountains and flat ice plains extending to Pluto's horizon.

# HOW DO WE DETECT THE PRESENCE OF VERY DISTANT PLANETS?

As we wander out into the rest of the Milky Way galaxy, we begin to realize how many other planets there are that circle other stars. To detect these exoplanets, we either require very sensitive measurements of the pull of (usually) a giant planet on the star it circles, or we must watch for regularly occurring drops in the amount of light coming from a star, which tells us that something keeps passing in front of it.

These exoplanet-hunting efforts were pushed into overdrive with the launch of the Kepler satellite in 2009. Kepler's primary mission was to watch for incredibly faint, regular drops in the brightness of more than 150,000 stars in the Milky Way. Its mission successor, TESS, launched in 2018, has taken up this hunt with a wider area of sky. These spacecraft mark a turning point in what has been a long, arduous hunt for these planetary relatives of ours—the first tentative discoveries, later confirmed to be real exoplanets, were in the late 1980s.

These early discoveries were the result of slow, painstaking measurements, produced by focusing on individual stars for long stretches of time, to be sure that a planet was the best explanation for whatever signal detected in the light. Kepler allowed us to monitor the stars in a much more efficient way, since it could watch for flickers in the starlight from many more stars at once. The number of known planets outside our solar system has accordingly skyrocketed.

There are two main methods of finding an exoplanet. The first involves catching the gravitational signature of a planet around the star you're interested in. This is easiest to do if you have a particularly massive planet circling your star, because in that case, the planet and the star are both orbiting their mutual center of mass. The center of mass can be thought of as the average

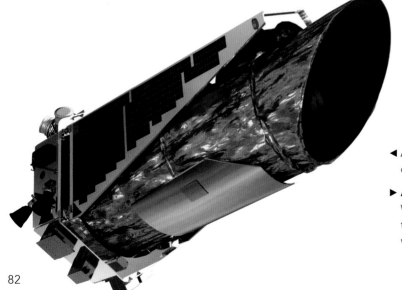

◀ A rendering of the Kepler telescope, which observed a tiny patch of sky for 9.6 years.

▶ **A small window onto the galaxy**
Within the Milky Way, the region of space that Kepler was able to survey is marked with a yellow cone.

position of all the atoms involved in the gravitational system. If the star is on its own, then its center of mass is precisely at its own center. If you add a planet into the mixture, the positions of all the atoms that make up the planet will start to pull the average position away from the center of the star, which still holds the majority of the atoms. If you have a large enough planet, then the center of mass will be pulled far enough away from the center of the star that the star no longer appears to spin in place, but rotates wildly, like a set of keys spun on your finger. We can spot this wobbling motion because it causes very specific features within the light produced by that star to move along with its own motion. In order to capture the motion of the star this way, you have to consistently watch the star with a special instrument. Because the wobble is more extreme with the most massive planets, this method tends to find that type of planet more than Earth-mass planetary cousins.

The other method for finding exoplanets is to wait for the planet to pass in front of the star, blocking out some of the star's light. You need a very sensitive camera, because the changes you're interested in are fractions of a per cent dimmings as the planet crosses the star. With enough time,

you can watch for regularly scheduled dimming of the same star, which flags up that you might have spotted a planet. This method finds large planets most easily. Large doesn't necessarily mean the most massive—a large fluffy planet works just as well as a large massive planet, but in general they do tend to track each other relatively well. To do this properly, you need to be taking regular measurements of individual stars over a long period of time, so that you can catch the repeated flickering of the star's light as it's blocked. This is exactly what the Kepler satellite did. Kepler had one of the most fantastically precise cameras ever produced, able to detect changes to the brightness of a star's light to much greater detail than ever possible from Earth, and its camera precision in combination with the efficiency with which it could observe the sky is what led to our exoplanet explosion. It found 2,662 confirmed exoplanets, and several thousand more planet candidates that may yet be confirmed with more observations. Kepler hunted in a very small area of the sky (see the diagram below), so there are many more planetary cousins yet to be discovered.

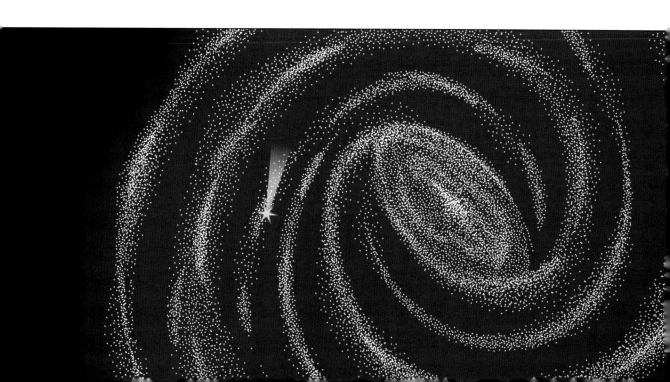

# WHAT DOES A PLANET NEED TO SUPPORT LIFE?

With so many planets out there, the opportunities for life, even within our own galaxy, seem almost endless. But getting the set of conditions to come together that have been required for life to flourish on Earth is a considerably more challenging task than just having enough planets to work with.

On Earth, the biggest requirement for life is water. Our planet is very good at growing things in every possible location, so long as it's near or in liquid water. Life arose extremely quickly after the formation of Earth (only about 200 million years after the impact that formed the Moon), which seems to indicate that once Earth had a surface with liquid water on it, there were not a lot of other stumbling blocks to overcome before life could spring forth.

Liquid water itself is a challenge. It usually means that the planet must be in a relatively narrow distance window away from its star, and have a surface upon which the water can rest. Effectively, we need rocky planets at exactly the right distance from their sun, such that all the water doesn't freeze solid or evaporate away. Outside of that distance band in a solar system that allows for liquid water (called the habitable zone), there are precious few opportunities for liquid water to exist (see the box, opposite).

There are, of course, a few exceptions, like Enceladus, a moon of Saturn, and Europa, a moon of Jupiter. Both of these moons are thought to have an ocean of liquid water under their icy surfaces, maintained entirely without help from the Sun. These small moons can maintain liquid water because the tidal forces from the massive planets they orbit are constantly stretching the rock at the cores of the moons. This stretching heats up the rock, and that heating provides the energy required to maintain liquid water. As a result, planetary scientists are very excited about the possibility of life on Europa and Enceladus, but in order to check, we'll have to send a craft to those moons to (very carefully) go and look directly.

For our exoplanet cousins, this requirement of liquid water means that of all the thousands of planets that have been discovered, we're most interested in the rocky planets that sit in this magic distance window from their star where liquid water can exist. These planets are extremely hard to detect, and push the boundaries of the sensitivity of our telescopes. It seems, from the Kepler data that's been studied so far, that about 20 per cent of all stars like our Sun have a rocky planet near enough to the star to have liquid water.

Proving that liquid water does in fact exist on those planets is more difficult still—you have to detect the signature of water in the atmosphere of a planet that is light-years away. Proving the existence of life will be an even more difficult task, but once we begin to find lots of planets with liquid water on their surfaces, the odds are pretty good that one of them will contain life of some form. It will of course be much easier to search for life within our own solar system, since we can actually go to these places and see what's there directly.

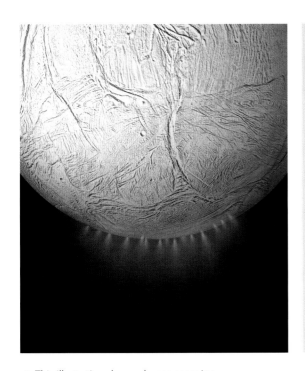

## WATER CARRIERS

It isn't correct to say that there's no water elsewhere in our solar system—it's just not present in a liquid form. There is a class of asteroids that are laden with water ice, behaving as an intermediate between the metallic asteroids and the primarily ice comets. We have also found water ice in a number of permanently shadowed craters around our solar system. The permanent shade guarantees that the temperatures never rise high enough to evaporate the water to outer space. Any of these water-laden objects, if they were to come barreling into a planet or a growing proto-planet, could deliver water, even if little had been present before.

▲ This illustration shows plumes escaping from fractures in Enceladus' crust, as we expect they might be seen. These plumes carry water, ice, and complex molecules out into space.

▶ **Europa's mysterious interior**
The current expected interior of the ice moon Europa, which orbits Jupiter, is a metallic core, surrounded by a shell of rock, and then a watery ocean, frozen over entirely with an ice crust.

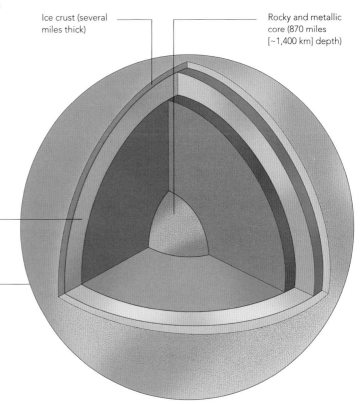

Ice crust (several miles thick)

Rocky and metallic core (870 miles [~1,400 km] depth)

Water ocean (~ 62 miles [100 km] thick)

Total radius: 970 miles (1,561 km)—approximately the same as the Moon

THE SOLAR SYSTEM

# Where can we look for life?

We often think about life emerging in shallow seas of warm water because that's the story of life on Earth. However, to get a sense of how many places outside our balmy shores might support life, we should probably explore Earth's oddest and least-hospitable corners first, from the nearly frozen waters in Antarctica to caustic caves and the hot mud of Yellowstone. Should we expect to find life only in sunlit, shallow oceans, or is an ice-encrusted salty ocean enough? And could the atmosphere of an otherwise inhospitable planet host life?

While we don't expect to find an analog to Yellowstone on another world, pushing at the boundaries of where life can exist on our planet helps us understand how charitable the conditions on another world would need to be before we rule them out as a candidate for some kind of life. In attempting to answer our question, we'll consider how life endures in a few of Earth's most extreme environments, and then look to some of our neighbors in the solar system to see how they compare.

## 1. Earth's irradiated upper atmosphere

While we know that the oceans and the surface of our planet are replete with living things, even the high reaches of Earth's atmosphere aren't off-limits to small forms of life. We recently discovered that our own atmosphere seems to be rather full of tiny microbes, suspended surprisingly far up above us. A hurricane scouting plane took samples of the air in 2013 a little over 6 miles (10 kilometers) above the surface, and found a phenomenal density of bacteria and fungi apparently thriving up there! At the very least, the bacteria they found were not all dead, which is a good step toward thriving. This discovery was a surprise, because the higher up you go in the atmosphere, the less protection you have from the high-energy ultraviolet (UV) radiation from our Sun. UV radiation is generally dangerous to life because it is high enough in energy to ionize atoms and molecules, meaning it kicks electrons out of their otherwise stable orbits; this can damage cells, causing them to mutate or die, depending on the severity of the damage. In humans this can lead to skin cells reproducing much faster than they should—it's one of the triggers for skin cancer. The atmosphere does a fairly good job of blocking most UV light, but the farther up you go, the less protection there is. 6.8 miles (10.9 kilometers) above the surface, 75 per cent of the mass of the atmosphere is below you, so this really is a very extreme, unprotected place for bacteria to survive. To find a large volume of bacteria, alive, seemingly unaffected by the UV dosage 6 miles up really was unexpected. At the moment, we think that Earth's storms are responsible for flinging so many bacteria nearly to the stratosphere, but it's the tiny mass of the bacteria that allows them to stay suspended up there. Finding live bacteria in our own atmosphere means that it's not unreasonable to suggest that the same thing might happen in other atmospheres.

Suspicions should immediately turn to Venus, everyone's favorite 860°F (460°C), runaway greenhouse, volcano-ridden, battery-acid raining planet. Now that description, while accurate, does not paint a picture of a particularly habitable place. We don't, in fact, expect to find life at the surface, where we have lost each and every one of our probes to a combination of crushing and melting after only a couple of hours. However, if you stay away from the surface, there's a layer in Venus' extraordinarily dense clouds that is positively balmy in temperature. It sits roughly 40 miles (64 kilometers) above the surface with a pressure about equal to that at the surface of Earth, and is around standard room temperature. Unfortunately for humans, this is also the part of the atmosphere of Venus that rains sulfuric acid. This toxic acid rain evaporates before it hits the surface, leaving a catastrophic layer in the atmosphere where no humans would dare to pass.

▼ The layers of Earth's atmosphere, to scale. Most clouds, and airplane travel, are in the lower portion of the atmosphere. Meteors are seen higher up, and above them are the aurorae. The International Space Station (ISS), in low earth orbit, cruises well above most of the atmosphere.

373 miles (600 km) upper limit of Earth's atmosphere

217 miles (350 km) International Space Station

The official start point of space (the Kármán line)

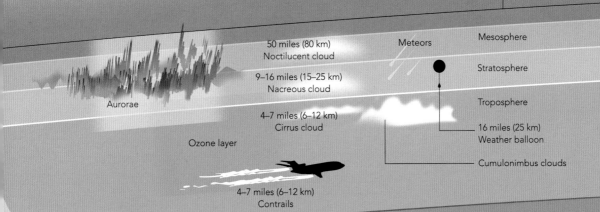

50 miles (80 km) Noctilucent cloud

9–16 miles (15–25 km) Nacreous cloud

4–7 miles (6–12 km) Cirrus cloud

Aurorae

Ozone layer

Meteors

Mesosphere

Stratosphere

Troposphere

16 miles (25 km) Weather balloon

Cumulonimbus clouds

4–7 miles (6–12 km) Contrails

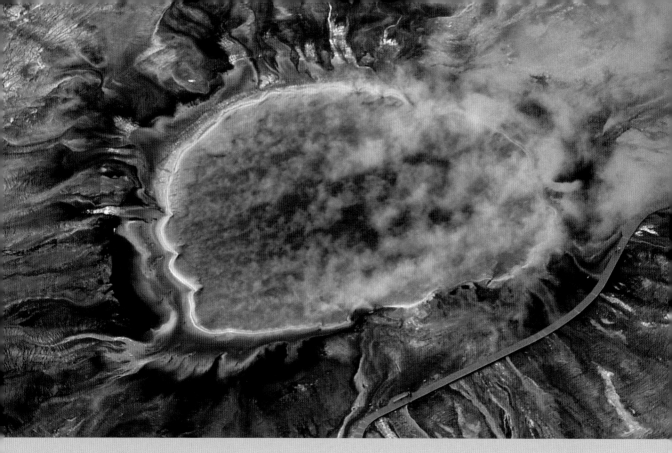

## 2. Toxic sulfur caves

For bacteria, however, the sulfuric acid in Venus' atmosphere may not spell immediate demise. Bacteria have been found on Earth that thrive in the most unexpected places, and some of them are fine with sulfuric acid. These particular organisms live in caves, form stringy mats, eat sulfur compounds, and produce sulfuric acid as a by-product.

They hang from the ceiling of the caves and are called snottites, or, if you prefer, snoticles. These caves are seriously unhealthy places for humans, because of a general lack of oxygen and also the sulfuric acid dripping from the ceiling (generally explorers need to be wearing heavy-duty protective gear and gas masks). But if similar bacteria were present (ignoring how they got there) in the cloud layer of Venus' atmosphere (with its reasonable temperature and pressure), they might be able to survive fairly well without worrying too much about the omnipresent sulfuric acid.

▲ An aerial view of the Grand Prismatic Spring in Yellowstone National Park. The color gradient seen in this hot spring shows the transition from truly sterile deep blue water to the domain of bacterial mats, here a vivid orange.

This sort of atmospheric thinking isn't limited to Venus—it's just that it's the closest planet in size to us, we have the most information on it, and it's probably the easiest to explore. Jupiter has also been subjected to the same thought experiments. There's a long line of science fiction authors musing on the idea of Jovians, many of them exploring the idea of creatures living in the clouds. Though it's extremely unlikely that there are airborne jellyfish or cloud whales on Jupiter, it's more possible that if life is hanging around there, it might be microscopic life, suspended in the more hospitable cloud layers. If we ever do find signatures of life hiding out in the cloud layers of our planetary siblings, that will be a strong hint that our exoplanet cousins might be hosts to similar forms of life.

## 3. Regions of extreme cold

A planet's atmosphere may turn out to be one reasonable location for life, but there's one more place we might find it—in stars that missed the amount of mass you need in order to start fusion burning in their cores: brown dwarfs (see pages 124–126). The coldest of them are really quite cold: the most extreme surface temperature is somewhere between –54°F and 9°F (–48°C and –12°C), which is right about the limit of the coldest survivable conditions for extreme-loving bacteria on Earth. Not all brown dwarfs would be suitable; they would need to be as Jupiterlike as possible, which happens only with the smallest of them, where the boundary between a Jupiterlike planet and a failed star is the fuzziest. But given what we know about stellar atmospheres today, if life could thrive in the high atmosphere of gas giants like Jupiter, then the lowest-mass stars, which may yet outnumber stars like our own, could be the home of starborne life.

## Verdict

Of course, this will remain a thought experiment until we're able to see for ourselves. There are missions that have been designed with present technology to look for life in Venus' clouds. There are others to check the positively hospitable-sounding warm oceans of the ice moon Europa. We may find that life among our Earth's siblings is not so rare as we had thought.

▼ **The Venusian clouds**
At the surface of the planet, the pressure is about 90 times more intense than on Earth, and a blistering 460°C (860°F). However, as you go up in the atmosphere, the temperature descends, and at the height where atmospheric pressure is "normal" for Earth (31 miles, or 50 kilometers), temperatures are between boiling and freezing.

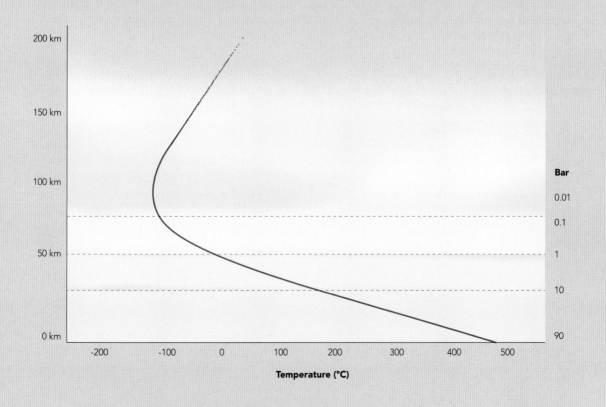

# HOW CAN WE SEARCH FOR LIFE ON OTHER PLANETS IF WE'RE LOOKING BACK IN TIME?

While we are indeed limited to seeing planets as they were when the light left them, our current time delay is relatively small. Even with its finely honed systems, the planet-hunting satellite Kepler only detected Earthlike planets up to 3,000 light-years away. If we were able to patrol our entire galaxy, our Milky Way is about 50,000 light-years from center to edge (so about 100,000 light-years across), and the most distant planet yet discovered is in the next nearest large galaxy, Andromeda, sitting about 2.5 million light-years away from us.

While we expect to find a planet around pretty much every star in our galaxy, we haven't been able to survey even a sizable fraction of the Milky Way, let alone the stars in Andromeda, which would be exponentially more difficult to observe. The farthest solid detection of an exoplanet through any means is still only about 21,000 light-years away, well within our own galaxy. However, the limitations of the speed of light mean that any images we get of exoplanets are just as out of date as they are distant from us. A planet that we see at 10,000 light-years distance from us will be an image that has traveled for 10,000 years.

On a geological timescale, 10,000 years is just a blip of time, but in a human terms, that time period has made a big difference. 10,000 years puts us back into the Neolithic era—the end of the Stone Age, around the time when pottery was developing and we were beginning to cultivate plants for agriculture. So an intelligent civilization, 10,000 light-years distant, that is just now looking for other life in the universe would spy Earth as a rocky planet, far enough away from the Sun that water could exist in its atmosphere. If they examined that atmosphere, they would notice that it is mostly nitrogen, with some oxygen and carbon dioxide in it as well, and that it contains water vapor. They would not be able to tell that there are creatures on that planet that are

10,000 years away from developing the Internet, neurosurgery, and machines able to detect tiny distortions in space itself.

This kind of time delay is one of the reasons that scientists get extra excited when they find a nearby rocky planet that might be able to have liquid water on its surface. If the planet is close to us, the time delay isn't as bad as a more distant planet. (Their excitement is also because it is much easier to observe these nearby planets in any degree of detail.) We only managed to detect the contents of the atmosphere of a slightly-bigger-than-Earth planet for the first time in February of 2016. Unfortunately, that planet is totally devoid of water, having an atmosphere of mostly hydrogen and helium, with some hydrogen cyanide thrown in for extra poisonous flavor. This planet is only 60 light-years away, so that image of it is only out of date as far as 1956. This particular planet won't have evolved into a friendlier, life-hosting planet in such a short time.

The farther out you go, the worse this time delay gets. In practical terms, we're unlikely to be able to do anything, be it communicate with, visit, or otherwise signal to a very distant civilization, if we did detect some signs of life on a distant exoplanet. The speed of light is the very best we can do—if we were to beam a radio message to a distant world, that radio wave would speed outward from Earth at precisely the speed of light, and no faster (because radio waves are a form of light). The speed of light is a fundamental speed limit within our universe—no information travels faster than the rate at which light could traverse the distance.

The same is true of any spacefaring civilization many billions of light-years away from us. If a super-intelligent civilization out there could build an impossibly large telescope, and had the power (and the time) to detect planets orbiting stars in a distant galaxy, and they happened to point it at Earth, they might see a very different world than the one we currently inhabit. If they happened to be 2.5 billion light-years distant, our planet's atmosphere would be in the middle of a dramatic change. This was when Earth was undergoing the "oxygen catastrophe"—the earliest photosynthetic bacteria were dumping oxygen into the atmosphere faster than it could be absorbed, and oxygen was slowly building up. As oxygen was a toxic by-product to the single-celled life that had been living in a

**LIGHT-YEARS**
A light-year is the distance that light, traveling at about 670 million mph (around 1 billion kph), can traverse in one year. One light-year is therefore 5.9 trillion miles (9.5 trillion kilometers).

delightfully oxygen-free environment, they would have had to adapt or die off. Observations of our planet from that distance would be able only to tell the observer that our planet existed, it had water in its atmosphere, and that we journeyed around our star every 365 days. There wouldn't be so much as a hint of our space-exploring future.

▼ The Andromeda galaxy is seen here through the eyes of the GALEX telescope, which observes the ultraviolet light produced by the hottest and most massive young stars.

# WHAT SHOULD WE DO IF WE FIND SIGNS OF EXTRATERRESTRIAL LIFE?

The unlikeliness of encountering life on one of our exoplanetary cousins hasn't kept us from setting up guidelines for what to do if it were to happen. There are laws to guide the behavior of nations in outer space, called International Space Law, which include: "Do not put nuclear weapons in space"; "Your country cannot claim a piece of space"; "You cannot build a military base on the Moon"; and "Space is for all countries." But what should we do if we discover life on another planet?

Most of the guidelines for what to do in case we find or are found by alien life have a similar set of instructions, whether in the SETI (Search for ExtraTerrestrial Intelligence) Institution's version, entitled "Protocols for an ETI Signal Detection," or the "Declaration of Principles Concerning Activities Following the Detection of Extraterrestrial Intelligence," which has been signed off on by a whole slew of administrative bodies. Simply put, they've agreed to be guided by the guidelines, which go roughly like this:

### 1. Triple-check your findings

If you think you have found a signal from extraterrestrial life, please make extra super-duper sure that what you think is an ET signal doesn't have any other more plausible explanation.

The discovery of the star KIC 8462852 (also known as Boyajian's star) is a fine example of what can happen when we jump to conclusions. Observations showed that the amount of light reaching Earth from the star was extremely inconsistent. It was rapidly dubbed the "alien megastructure star," after the suggestion that a large, artificial structure of extraterrestrial, intelligent origin might be blocking the light. The fanciful theory sparked a huge wave of excitement. After further observation, however, that star is now thought to have a very thick disk of extremely fine dust orbiting it (which can block large fractions of the star's light), possibly in combination with a swarm of comets, which can account for the faster flickering that was seen. Even with a relatively rapid scientific turnaround away from the alien construction project, there were a huge number of articles on the possibility. I can only imagine the sort of media explosion and cultural whiplash we would have on our hands if we announced that we had discovered intelligent life.

### 2. Get a second opinion

Pass what you think to be a detection to other scientists, and let them check your work. This process might mean getting more data and observations, or just someone else verifying that they can reproduce your numbers. At this stage, please do not share your findings with the media. It may well be that what you have found, while very interesting, is not evidence of alien life. If a scientist were to announce that he or she had found aliens, only to be contradicted by evidence showing it to something astrophysically bizarre rather than aliens, there would be an instant loss of trust in that person. This second stage is to make sure that at the very least, reliable numbers and information are announced.

### 3. Notify the UN

If the signal has not gone away and still looks like alien life, even after some serious checking, you should tell the Secretary General of the UN and a whole string of scientific bodies, which will let the rest of the scientific community know. You should also swiftly tell the rest of the world. But hey—if it was your data that did the discovering, you get the honor of hosting a very high-profile press conference, and will probably spend the next year and a half (at least) answering questions.

### 4. Go public

After that, it's time to go public. However you detected intelligent aliens, please tell other scientists so we can do more of it. This means explaining what you did at conferences and probably making your data public. Also, please put your data in as many places as possible. The last thing we would want at this stage is to lose it. Put it in 10 places. Put it everywhere. Please do not lose the evidence of aliens to a hard drive failure. If the aliens were detected on a particular frequency (say we heard a radio transmission of theirs), that frequency should probably be protected so we can keep listening without too much interference. Radio-quiet areas are pretty rare nowadays, and a lot of things can cause radio waves (a bizarre chirp detected by a radio-sensitive telescope was recently proven to be generated by someone opening a microwave door).

### 5. Don't make contact

You do not get to write back to the aliens, not without first clearing it with majority of the planet, anyhow. At that point, contacting aliens has an impact on the whole human race, not just science, and not just the country that the scientist who discovered it lives in. The decision of what to do next should be discussed by everyone.

But because these are only guidelines, if some private company happened to do the discovering, and disregarded all of the above rules, the way things sit currently they would be doing nothing illegal. The guidelines are also unlikely to be needed for many centuries.

▶ Radio telescopes such as this have been used for searches for signals of extraterrestrial life. So far, we've found nothing convincing.

## A QUESTION OF SCALE

The unlikeliness of making this kind of discovery comes from the fact that our galaxy is a very, very large place. Our planet is supremely unremarkable in terms of the star we orbit, our location in the galaxy, and our galaxy's location, so it seems likely a similar configuration should have produced itself elsewhere. The trouble lies in the vastness of how many "elsewhere" options there are.

Astronomers traditionally tackle the unlikeliness of finding alien life by means of the Drake equation (see the box on the right), which takes all the pieces we think need to align for life to evolve, and multiplies together all the probabilities that they'll occur in the same place, at the same time, within our galaxy. There are at least 100 billion stars in our galaxy, and hundreds of billions of galaxies in the universe. As noted earlier, virtually every star we've looked at in our galaxy has at least one planet, which means we're dealing with an overwhelmingly large number of planets in the universe. Earth went from having liquid surface water to supporting life in about 125 million years, a cosmic blink of an eye—that doesn't seem to be the hard part. The tricky part is getting from bacteria to a species that contemplates the skies above. But even if you're extremely pessimistic about the fraction of planets that have any form of life, and pessimistic about the fraction of planets with intelligent life, the sheer number of planets out there dominates these calculations. There simply has to be other life out there, and if not in our galaxy, in another.

However, the Milky Way is huge. As we saw on pages 90–91, it takes light 50,000 years to travel from the center to the edge of the galaxy, and 100,000 years to make it all the way across. And the distances between galaxies are even more extreme. Light from the Andromeda galaxy takes 2.5 million years to reach us, and that's our nearest neighbor. 2.5 million years ago, humans were only at *Homo habilis*, mastering early stone tools. We don't have any record of modern humans from

### THE DRAKE EQUATION

In a simple form, the Drake equation asks the following series of questions:

- From the number of stars in our galaxy, what fraction of those stars will have planets?
- On average, how many planets are at just the right distance from their star, where liquid water can exist at the surface?
- Of the planets with liquid water hanging around, how many of those should we expect to have any form of life, no matter how simple?
- What fraction of those planets with life will also have intelligent life?
- And what fraction of those intelligent life forms are still around now and potentially able to communicate with us?

earlier than 200,000 years ago. Our first radio telescope was built all of 80 years ago.

We can do a few more rough calculations to figure out how bad the problem is. Let's assume that you work out that by optimistic numbers, there should be 15 intelligent, communicable civilizations just in our galaxy. If those civilizations are randomly scattered around the galaxy, they're separated by about 23,000 light-years. If there's only one civilization per galaxy, you're back to separations of millions of light-years. Communication between civilizations (even in the most optimistic of cases, where we send a signal out at the speed of light) would be impossible.

▶ An illustration of Kepler-186f. It is the first Earth-mass planet confirmed to exist in the habitable zone around its star. Drawn here as a watery world, it's unknown if this kind of surface actually exists.

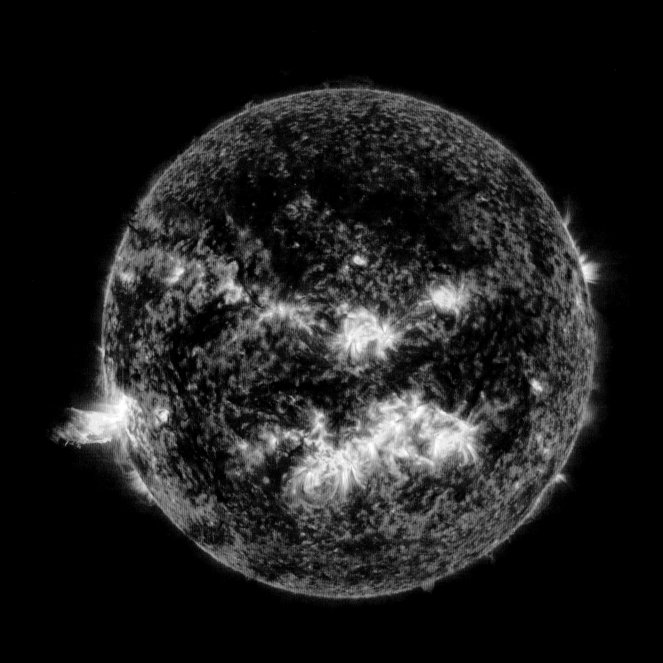

# 4

# STARS

A glance at the night sky tells us that our Sun is by no means the only star. Fortunately for us, our star is fairly average in most ways, and we can learn about stars generally by studying ours in detail. From there we expand outward: How different can a star be from our own?

# HOW DOES OUR STAR PRODUCE SUNLIGHT?

Stars are tremendous collections of gas, compressed into such a high density that the temperatures at their cores allow for the construction of new elements in their centers. For all the endless fascination of our neighboring planets, it's the stars that make planets possible. Without them, life on our planet would certainly not exist, as the sunlight that warms our planet enough for liquid water is the after-effect of the Sun's own construction of these new elements—and the Sun is of course a star.

If we humans are the children of Earth, the Sun is our grandparent. Our planets would not have formed except around a star. The left-over remnants of the gas cloud that collapsed to form our Sun went into forming the planets, as we saw in the last chapter. Without the initial collapse of the gas to form the Sun, there would have been no dense gas and dust ready to form Earth and its sibling planets. In the way of many grandparents, the Sun fed our planet and its life-form grandchildren—in our case, a rich diet of sunlight for a watery world, allowing photosynthetic plant life, and more complex animal life in turn, to flourish.

## THE SECRET OF SUNLIGHT

So where does the sunlight emanating from Earth's parent star come from? To answer that we're going to have to dive into the center of the star, through a number of layers of stellar material to the core, where it all begins.

Our Sun, known formally as Sol, is a pretty good prototype for how stars are constructed and how they behave; it is a middle-aged star, in the middle of the longest period of stability it will ever have, and it is not particularly large or small when we compare it to the rest of the stars in our galactic neighborhood. As a result, its interior is also pretty average, and a journey into its core will give us some insight into how most stars work.

> Stars are mostly made of hydrogen gas, largely out of sheer convenience—hydrogen is the most abundant gas in the universe.

Stars are mostly made of hydrogen gas, largely out of sheer convenience—hydrogen is the most abundant gas in the universe. If you collapse any random patch of gas in the universe, you're likely to have gotten yourself a big pile of hydrogen, perhaps with a few other elements tossed into the mix in small quantities. Our star is no exception; it's mostly hydrogen plasma. Plasma is a phase of matter one step more extreme than a gas, where the electrons that normally make up an atom have gained so much energy that they've been able to vault away from their nucleus, and are wandering freely. The nucleus of the hydrogen atom, now that its electron has left, remains as a simple proton (see the box on page 100).

As we journey to the core of the star, the hydrogen is subject to increasing temperature and pressure. A star is a very precise balancing act between the

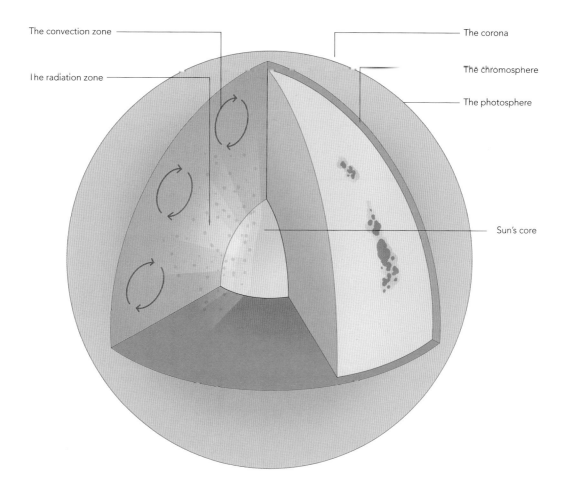

The convection zone

The radiation zone

The corona

The chromosphere

The photosphere

Sun's core

▲ **The structure of the Sun**

All the Sun's light is produced in the central core. From there, it travels outward to the radiation zone, taking more than 150,000 years to bounce randomly through the Sun. When it happens to travel far enough outward, it will find itself in the convection zone, where high-temperature plasma bubbles to the surface. What we typically call the "surface" of the Sun is formally known as the photosphere, the point at which light is free to leave the Sun unhindered. The chromosphere and the corona are two components of the stellar atmosphere, a diffuse envelope of very hot gas that is most easily visible during a solar eclipse.

STARS

crushing force of gravity, pulling all particles inward, and a resistant pressure that comes from both heat and a reluctance of particles to be pressed too close together, which forces electrons and protons farther apart from each other.

In the very center of the star, the heat and pressure are sufficiently high that the hydrogen nuclei slam into one another at such high energies that they can stick together. This is nuclear fusion. It's a gradual process that involves the release of light and the conversion of protons into neutrons (see the box on the right). Ultimately, you find yourself with a fully constructed helium nucleus: two protons and two neutrons, all stuck together. This smashing together of hydrogen and helium nuclei can, in principle, be used to build more complex atoms, but the heat and pressure required to build those complex atoms is higher than we see in the Sun. For a star the size of our Sun, the building of hydrogen into helium is as complex as it gets.

This fusion process only occurs in the dense core of the Sun, which extends from the very center out about one fifth of the way to the surface. This means the core is 172,915 miles (278,280 kilometers) wide, from edge to edge, which is large, but nothing when compared to the 864,576 miles (1,391,400 kilometers) that the Sun itself spans. Every single beam of light that the Sun produces comes from this small region in the very center of our grandparent star.

For a beam of light, escaping the core of the Sun is not a difficult task; this region is totally transparent to light. Two invincible friends sitting in the Sun's core would have no more trouble seeing each other than they would through thin air. It's the middle zone that serves as a roadblock for light: the radiation zone (see pages 102–103).

## PROTONS AND NEUTRONS

A proton is a component of an atom, which is found in the core of an element. The number of protons tells us what element the atom is. Protons are positively charged. The simplest atom, hydrogen, is a single proton and a single electron.

Neutrons are also critical components of an atom. They have no electric charge. Changing the number of neutrons in an atom changes the isotope of the atom, but does not change the element.

Two protons (the nucleus of hydrogen), when slammed together with enough energy can react to create a neutrino (high-energy particle), which will escape the Sun quickly, and a positron (like an electron, but with positive charge), creating a heavy hydrogen nucleus of a proton and a neutron together.

▼ **Creating a helium atom out of hydrogen**
Rather than being a process that requires two hydrogen atoms to build a single helium atom, six hydrogen atoms are required to build the two protons and two neutrons of a hydrogen nucleus. Two of the hydrogen nuclei are returned to the solar environment and can be used as part of another helium-building process.

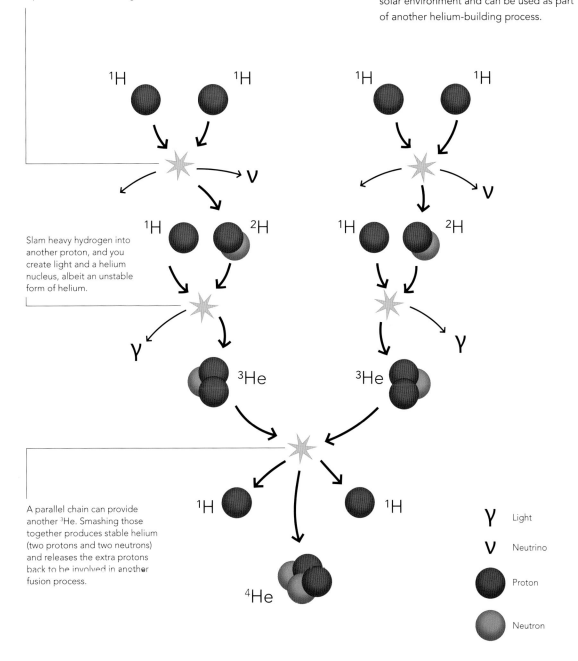

Slam heavy hydrogen into another proton, and you create light and a helium nucleus, albeit an unstable form of helium.

A parallel chain can provide another $^3$He. Smashing those together produces stable helium (two protons and two neutrons) and releases the extra protons back to be involved in another fusion process.

$\gamma$  Light

$\nu$  Neutrino

Proton

Neutron

# WHAT IS THE RADIATION ZONE?

The radiation zone is the area surrounding the core of the Sun, where energy is moved through radiation. Radiation itself is something we usually think of in terms of cancer treatments or nuclear waste. Fundamentally, it involves light (photons) serving as carriers for energy; high-energy photons leave a very energetic area and travel to a less energetic area, bumping into something and depositing their excess energy. This process then evens out the energy levels everywhere.

Closer to home, this is the method behind a conventional oven, or a sunburn, or holding your hand over a hot surface to see if it's safe to touch. For a sunburn, a photon of light has left the Sun, traveled 93 million miles (149.5 million kilometers) through space, through Earth's atmosphere, and bumped into your skin. Your skin absorbs the energy that the sunlight carried, and given enough time and enough absorptions, your skin will begin to protest at this energy dumping, which is damaging its cells. Living cells do not like radiation much, which is why we humans depend so thoroughly on the protection of our atmosphere to survive.

Light is caught by these same rules in the Sun's radiation zone. Each photon moves from the hot core, bounces into some of the hot plasma in the radiation zone, and gets absorbed. What we have is a stream of newly formed light coming from the fusion reactions happening in the core, ready to flow out of the Sun, which then hits the boundary of the radiation zone, only to be absorbed by the material there. But it's hot there, near the core of the Sun, so the energy is quickly spat back out in the form of a photon, going in a new random direction. That photon doesn't have far to go before it runs into a new plasma particle, and gets absorbed and then spat out once more, again in a random direction. Any photon we could choose to follow will get trapped in this cycle of absorption followed by emission in a random direction.

Because the material here is very dense, the photons never have far to travel before their next bounce, and each time they hit something new, they're redirected in a random direction. This directionless motion is called a random walk, and if you have a goal in mind, it is one of the least effective ways of getting there (as shown in the illustration opposite).

For a photon of light trying to escape the Sun and perhaps eventually land on a planet, the radiation zone is a trap that lasts for at least 100,000 years.

This directionless motion is called a random walk, and if you have a goal in mind, it is one of the least effective ways of getting there.

▶ **25,000 steps of a random walk**
At each step, the particle being traced randomly chooses whether to go up, down, right, or left. This results in a complex pathway, and is a remarkably inefficient method of crossing a fixed space.

# WHAT HAPPENS IN THE CONVECTION ZONE?

Let's say that, after 200,000 years of effort, a lucky photon has escaped the Sun's radiation zone (see pages 102–103). Up next for this intrepid adventurer is the convection zone. Convection, like radiation, is a method of moving energy around, but instead of doing everything through a photon pack mule, convection moves the particles around themselves. If you want to mix hot and cold substances together, you're going to need convection to shuffle around the molecules.

The most common everyday example of convection is to look at a pot of boiling water, where the fluid in the pot is roiled and moving around with the bubbles rising to the surface (not just sitting still), and donating its energy to its surroundings.

## BUBBLING UP AND OUT

Convection takes over in the Sun when we've reached a sufficient distance from the core that everything has cooled down a bit, and so photons of light leaving the radiation zone at long last are absorbed by the plasma in the convection zone, and that material itself is heated by that absorption. The plasma begins to rise—like a bubble of water in our boiling pot—pushing directly toward the surface of the Sun. If you have a very powerful solar telescope, you can even see these bubbles on the surface of the Sun.

As the hot material rises, the plasma closer to the surface of the Sun, which is cooler, sinks back down to replace it. The bubbling up of hot material only takes about a week, which, in comparison to the 100,000 years (at least) it took for the light to escape the radiation zone, is positively instantaneous. Once the bubbles reach their peak, the light stored inside this heated material is free to leave the Sun entirely and stream outward into space in whatever direction that tiny piece of the star's surface happens to be pointing.

This point, where the Sun becomes transparent to light again, is the photosphere, and it's this surface that we see in images of the Sun. There is a tenuous, thin atmosphere beyond the photosphere, made of very hot gases (much like the rest of the Sun), but the best time to see that atmosphere is during a solar eclipse, when the bulk of the Sun's light is being intercepted by the Moon.

---

The plasma begins to rise—like a bubble of water in our boiling pot—pushing directly toward the surface of the Sun.

---

▶ The Daniel K. Inouye Solar Telescope produced the highest-resolution image of the Sun's surface ever taken. In this picture, we can see features as small as 18 miles (30 kilometers) in size for the first time ever. The cell-like structures, each about the size of Texas, are the signature of violent motions that transport heat from the inside of the Sun to its surface. Hot solar material (plasma) rises in the bright centers of "cells," cools off, and then sinks below the surface in dark lanes in a process known as convection.

# HOW DO THE MAGNETIC FIELDS OF EARTH AND THE SUN COMPARE?

We've seen the path that light takes, from its creation in the Sun's core to its eventual escape. However, that's not yet the complete story of our cosmic grandparent: what's missing is the magnetic field of the Sun. The Sun's magnetic field has nothing solid to anchor itself to and so it, like the roiling surface of the Sun, is an ever-changing entity, gradually tangling and twisting as years go by. These twists and knots produce sunspots, visible defects in the otherwise pristine solar surface. Sunspots can be the size of Earth and are observable from Earth with a solar telescope.

Earth's magnetic field is a bit more familiar to us; it orients our compasses, and it protects our planet from radiation that would be harmful to life, producing the aurora effect as some of these high-energy particles from space end up slamming into our atmosphere (see pages 20–21). It is pretty straightforward; it has a North and a South Pole, like a bar magnet, producing an orderly and neat protective bubble that surrounds Earth. It's like an overinflated donut that encompasses a wide swath of space around our planet. The Sun's magnetic field is not nearly so neat.

## MAGNETIC MOTION

There's a critical difference between the magnets we can stick to our fridges and the magnetic nature of a planet or a star, and it all comes down to the fact that refrigerator magnets are solid, and planetary cores are not. For refrigerator magnets, the magnetism is effectively frozen in place within the atoms of the magnet, and the magnetic field surrounding that magnet is totally static—it doesn't change over time, and no property of the magnet will change its strength or shape.

The magnetic fields of Earth and the Sun are very different beasts. On Earth, the magnetic field is formed by the motions of magnetic nickel and iron in the core of the planet, which generates an electric current. Because it's generated by the motion of material deep within Earth, the magnetic field is not nearly so static as a refridgerator magnet. Magnetic North, which is the location that compasses point to, drifts by a few degrees every year, and wanders around the general area of Geographic North (see the box opposite).

On the Sun, the magnetic field is also generated by the motions of material within its interior that produce an electric current. So you might think that the Sun's magnetism should behave like Earth's magnetic field, but there's a critical difference between Earth's behavior and the Sun's; a generational divide, if you like. Earth, very conveniently for us humans, has a solid surface. The solid surface means that the planet has to rotate as a coherent unit. There's no way for the Bahamas to have a shorter rotation time than Great Britain—both islands are fixed to the rock below, and both have a 24-hour day.

On the Sun, there's no rock to anchor the rotation speed of the star to a single value. As the entire contents of the Sun are a superheated plasma, it is relatively easy for something at the equator of the Sun to move faster than to something at the poles.

In fact, this is what we see. The equator of the Sun rotates a few days faster than the poles do. Unlike Earth, whose surface is mostly nonmagnetic, the Sun's surface is welded to its magnetic field, and so the rotation differences between poles and equator mean that the magnetic field gets pulled along into an increasingly twisted configuration. After about a month, when the Sun has done one full rotation, instead of having magnetic field lines that smoothly travel from pole to pole in the same kind of donut shape as with Earth, our star has twisted the middle of its magnetic field considerably to the side (see the diagram on page 108).

## GEOGRAPHIC NORTH

Geographic North is the physical point where no stars would rise or set, since you're exactly at the top of the spinning Earth. Magnetic North is the top end of the magnetic field, and is most often not at the same place.

▼ **Geographic and Magnetic North**
Geographic North is defined as the top of the rotation axis of planet Earth. Magnetic North, on the other hand, is defined by the magnetic field's location, and the magnetic field tends to drift slightly with time.

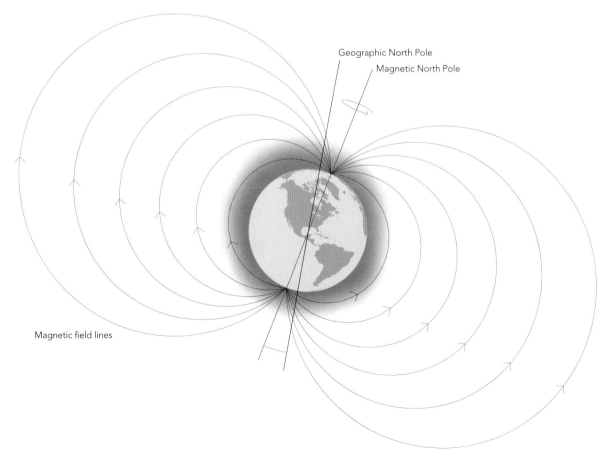

Geographic North Pole

Magnetic North Pole

Magnetic field lines

If you let the Sun continue to rotate (not that we have much choice in what the Sun does), these magnetic field lines will get more and more twisted over to one side, until they no longer resemble any kind of ordered pole-to-pole path. At this point, the tangled magnetic field is unstable enough that it will begin to pop little sub-loops of magnetism out of the surface of the Sun. The more repetitions of this month-long twisting process that go by, the more of these little loops there are, bursting out of the surface. These sub-loops are what give our solar grandparent its spots. Where the loops touch the surface of the Sun is where we see dark patches, known as sunspots.

## SUNSPOTS

Sunspots are slightly cooler regions on the surface of a star. While the density, pressure and temperature of the Sun's surface are normally consistent, the appearance of a magnetic field loop causes them to fluctuate. The loop presses the gas at the Sun's surface outward, allowing the gas in that region to cool down. The lower temperature of the sunspots make them appear darker than the surrounding area. As the Sun's magnetic field becomes increasingly tangled, more and more of these dark spots appear.

▼ **The Sun's magnetic field**
Over time, the magnetic field of the Sun becomes increasingly tangled, as the equatorial regions of the Sun rotate faster than the poles. Given enough time, the magnetic field will shift from looking vertically aligned, to something that is largely twisted sideways.

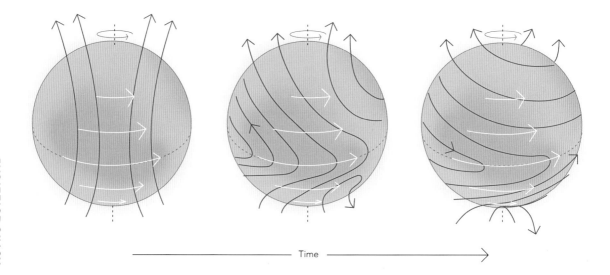

Time

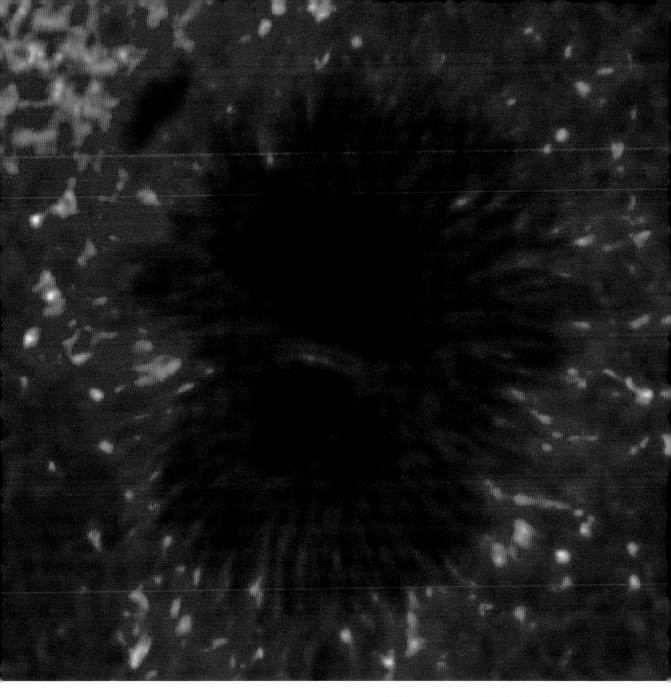

▲ A sunspot captured by the Hinode satellite. Launched in 2006, the satellite is operated by JAXA, and it nearly continuously observes the Sun.

# HOW DAMAGING ARE SOLAR STORMS TO EARTH?

Sunspots are not static objects on the surface of the Sun, and as the magnetic field of the star changes, their size and shape can change in response. The Sun's magnetic field has another trick to play as well: the sub-loops of magnetism aren't permanent.

If the magnetic field of the Sun is sufficiently tangled, those loops pushing out of the surface can themselves become twisted. If a loop crosses itself, the magnetic field will snap itself into a new, shorter configuration—where the loop connected to itself will be the new, shorter magnetic pathway. If the loop has crossed near the base, the magnetic field will create a very low, flat loop, just above the surface of the Sun, where previously there had been a more lengthy, extended loop. Above that, we have a newly freed strand.

### CORONAL MASS EJECTIONS

This free strand is made of protons, electrons, and other bits of the surface of the Sun, which up until the magnetic reconnect described above, had been able to flow smoothly along the magnetic loop. Without the loop's presence, the strand is no longer attached to the surface of the Sun. Nor will it stay hovering over the surface of the Sun for very long; a lot of energy is released at the moment that the magnetic field reconnects into a flatter loop, and that energy goes into violently kicking this strand outward, into the space surrounding the Sun.

If this reconnection event is large enough, the resulting blowout of the disconnected strand of the Sun is called a coronal mass ejection. Any strong burst often goes by the name "solar storm." Most of the time, Earth is well out of the way of such outbursts, and so the particles swing outward through the solar system without encountering

much resistance. If, however, the alignment of events is unlucky, sometimes Earth winds up directly in the path of one of these high-speed particle dumps. Earth's protective magnetic field takes the brunt of these blasts, so we humans on the surface of the planet are usually in no danger. Our orbiting electronics, on the other hand, are much more exposed to these kinds of phenomena.

We have put a lot of satellites in orbit around our planet, and every one is a very delicate set of electronics. If the Sun sends one of these blasts of material toward our satellites, suddenly the electronics within them are being bombarded with far more charge than they were designed to withstand.

---

Most of the time, Earth is well out of the way of such outbursts, and so the particles swing outward through the solar system without encountering much resistance.

---

▲ This image, acquired by the ESA-NASA *SOHO* spacecraft, was taken on January 8, 2002. It shows a widely spreading coronal mass ejection (CME) as it blasts more than 1 billion tons of matter out into space at millions of miles per hour. A second image, from the EIT disk imager on a different day was enlarged and superimposed on the C2 image so that it filled the occulting disk for effect.

The Sun sends a certain amount of protons and electrons our way every day, but the density involved in a coronal mass ejection is so high that satellites, if unprotected, can short-circuit. The constant bombardment of electrons means that any metal parts of the satellite can build up a charge on the satellite itself, like shuffling in your socks on carpet. If enough charge builds up, the satellite will short itself out, destroying critical components. This is mostly an issue for satellites in very high orbits (GPS satellites are one example), and that are less deeply embedded in the magnetic field of Earth.

The International Space Station (ISS), for instance, is in a low-enough orbit that it's mostly protected from these kinds of solar storms by the magnetic field, but every so often, there's a strong enough storm that the astronauts on the ISS take shelter in more strongly shielded portions of the structure, just for good measure.

## INCOMING

Solar storms and coronal mass ejections are almost never strong enough to press back against Earth's magnetic field with such intensity that electronics on the ground are affected, but it has happened. In 1989, there was a stupendously strong solar storm, which managed to affect the magnetic field of Earth to such an extent that power cables started to carry current in a way that they weren't designed for. A series of systems then failed to manage this unexpected current, and the failures led to 6 million people losing power in Quebec. In repairing the damage done in this event, the system has been redesigned so that it is harder for space weather to break it so easily, but even then the modern power grid is widely considered unprepared for another, similar event. This is one of many reasons to maintain a fleet of telescopes that monitor our star—if we can see these bursts when they leave the Sun, we'll have a few days of warning before they reach the orbit of Earth.

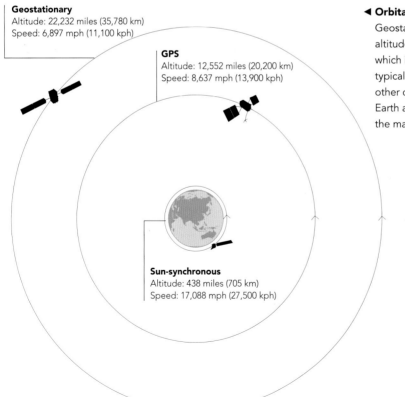

**Geostationary**
Altitude: 22,232 miles (35,780 km)
Speed: 6,897 mph (11,100 kph)

**GPS**
Altitude: 12,552 miles (20,200 km)
Speed: 8,637 mph (13,900 kph)

**Sun-synchronous**
Altitude: 438 miles (705 km)
Speed: 17,088 mph (27,500 kph)

◄ **Orbital altitudes**
Geostationary orbit is at a much higher altitude than the typical GPS satellite, which in turn is much higher than the typical low-earth orbit of the ISS and other crewed missions. The closer to Earth a craft is, the more protected by the magnetic field it will be.

These solar storms tend to concentrate in a few-year period when the Sun's magnetic field is at its most tangled. But there's always an end in sight—after the eleventh year of tangling up its magnetic field, the Sun resets itself. The magnetic field turns over and begins again as a smooth magnetic loop, from the top pole to the bottom pole. Once the magnetic field resets, the Sun goes quiet and sunspot-free for a few years while it gradually restarts the process of tangling itself up again. After another 11 years, there's another reset and turn-over, completing the cycle.

▲ A CME blasting off the Sun's surface in the direction of Earth. The left portion is composed of an EIT 304 image superimposed on a LASCO C2 coronagraph. Two to four days later, the CME cloud is shown striking and beginning to be mostly deflected around Earth's magnetosphere. The blue paths emanating from Earth's poles represent some of its magnetic field lines. The magnetic cloud of plasma can extend to 30 million miles (48.3 million kilometers) wide by the time it reaches Earth. These storms, which occur frequently, can disrupt communications and navigational equipment, damage satellites, and even cause blackouts.

# WHAT IS THE HELIOSPHERE?

A steady stream of particles blows outward from the Sun, forming what we know as the solar wind. This wind marks the interior of a bubble that encases our home, and all of our planetary siblings, bounded by the outer limits of the Sun's magnetic field. From our position inside this bubble, we don't feel the influence of the other stars in our galaxy. Here, in the solar system, space is relatively calm, aside from the occasional blast from the surface of our Sun (see pages 110–113).

This boundary between our Sun's regime and the rest of the galaxy is one way to define the edge of the solar system. We've given it a name: the heliosphere (from the Greek meaning "sun-sphere"). Beyond it, the direction of the cosmic wind changes, signaling a shift into what's called "interstellar space"—the space between sibling stars.

We're still learning about the nature of the edge of our solar system. We've only aimed a few spacecraft out that far, and to date, only *Voyager 1* and *Voyager 2* have crossed into interstellar space. *Voyager 1* was determined to have officially made it out of a transition region and reached the other side in August 2012. *Voyager 2*, its duplicate craft, was on a slightly different pathway out of the solar system and took four additional years to exit the solar system, even though they were only launched 16 days apart from each other.

▶ The twin *Voyager* spacecraft are currently the most distant human-made objects. The *Voyagers* were launched in 1977, and between the two craft, all of the gas giant planets were imaged. Uranus and Neptune have only ever been visited by *Voyager 2*.

## COSMIC RAYS

We once thought that *Voyager 1* would leave the solar system all at once, very much like stepping through the boundary of a soap bubble. Instead, *Voyager 1* encountered a much more gradual transition, where a few of the expected observations for leaving the solar system were seen, but not all of them at once. There was, however, a particularly dramatic moment on August 25, 2012, where the amount of plasma the spacecraft was plowing through increased sharply, as did the number of cosmic rays.

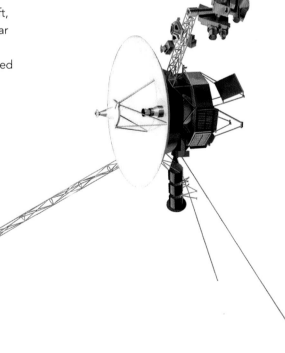

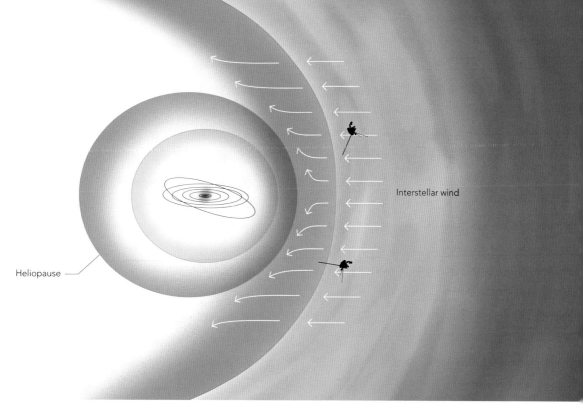

Heliopause

Interstellar wind

As we've seen, cosmic rays are very high-energy particles, which typically arrive into the solar system from elsewhere in the Milky Way galaxy. Our star shields us from many of these cosmic rays, so once *Voyager 1* had left the Sun's protective bubble, it made sense that it would have encountered many more. For *Voyager 1*, this transition occurred at around 121 astronomical units (au; see page 52), a distance that takes light itself just over 16 hours and 30 minutes to traverse. Compared to the vast distances to the nearest stars, which it takes light entire years to cross, even this large stellar bubble shrinks to a very local space.

With no more craft on the way to the edge of the solar system, other than *New Horizons* (see pages 80–81), and with a journey time of more than 10 years to get there, should we decide to go exploring the boundaries of our solar bubble once more, there will be a wait. Nonetheless, there's still useful information coming from *Voyager 1*.

**▲ Intrepid explorers**

The relative positions of the two *Voyager* spacecraft. Both have now left the enclosure of the solar system and are traveling through interstellar space.

*Voyager 2* crossed into interstellar space in November 2018, at 120 au from the Sun. With one more functional instrument, *Voyager 2* confirmed its exit from our solar system by recording a jump in the density of plasma surrounding it. From *Voyager 2*, we learned that this particular boundary of the solar system, marked by an increase in cosmic rays and plasma, is rather abrupt, and seems to be symmetrical; *Voyager* 1 and 2 took very different paths out of our solar system, and yet reached the edge at similar distances. Once we've reached the outer edges of our solar system, we've exhausted the region of space that our own Sun can influence, and it's time to turn to its siblings, its companions in the orbital dance around our galaxy, and the objects that illuminate our dark nights—the other stars.

STARS

# IN WHAT WAYS DO STARS DIFFER?

Our particular star is typical of its siblings, but that doesn't mean all the stars are identical copies of each other. An average human is of average height and weight, but humans are incredibly varied, beyond that average set of values. The stars in the night sky are equally diverse, ranging from stars tens of times more massive than our Sun, which glow a toxic, cancer-inducing blue color, to the faint red dwarfs, less than one tenth the size of the Sun, which can glow a reddish purple, not even as warm as the surface of Venus in some cases.

Our family of stars is collected into our galaxy, the Milky Way. Each of our stellar family members, along with any planets they have in their tow, orbits the center of the galaxy on its own path. Even with the huge numbers of stars in the Milky Way, the sheer volume of space in the galaxy means that no two stars will ever hit each other. Our galaxy is shaped in an almost two-dimensional disk—proportionally speaking, the Milky Way is much thinner than a sheet of paper. The family of stars that inhabits this disk was made possible by the galaxy's immense gravitational pull, which collected gas to itself in the early universe. Without that initial pulling together, it would be difficult to find the concentrations of gas that is needed to construct a star.

## MADE OF GAS AND DUST

Most of the stars in our galaxy formed, either in the Milky Way as we know it or at a much earlier time, before the galaxy was fully constructed. But it is this common pool of gas and dust that brings these stars together as family.

If we want to differentiate a stellar sibling from the rest of its family, there must be a critical set of information to know about that star. The simplest one is color (see pages 118–121).

Even with the huge numbers of stars in the Milky Way, the sheer volume of space in the galaxy means that no two stars will ever hit each other.

▶ Dark lanes of dust traverse the giant elliptical galaxy Centaurus A. In this image, young blue stars are visible, alongside the tell-tale pink glow of ionized hydrogen, which also traces the formation of new stars. In other images, Centaurus A has been determined to have a strong warp, likely to be the result of an ancient collision with another galaxy. Such collisions can trigger the creation of many clusters of new stars, such as those seen here. The dust itself is seen in silhouette against the swarm of brighter stars behind it. This image was taken in July 2010 with Hubble's Wide Field Camera 3.

# WHAT DETERMINES A STAR'S COLOR?

Stars produce light of many colors; our own star produces every color in the rainbow—very literally. But unless we go to some effort to split all these colors of light apart, either through a scientific instrument or through the remarkably effective prism of the raindrop, the light that reaches our eyeballs from our Sun appears white. This is because all of these colors of light have mixed together, and the average color is pretty much smack in the middle of the visible spectrum.

Some stars in the night sky are very obviously not white-light stars—Betelgeuse, one of the bright stars in Orion's shoulder, is a red supergiant with an orange glow even to the unaided eye. The mixture of light coming from this star has much less blue than our grandparent Sol. The lack of blue light being produced means that the red light can tilt the average color over toward a more reddish-orange.

We can dig a little deeper than this—rather than just looking at the average color coming from the star, we can break apart the light into its respective colors, and check exactly how much more red light is coming from a star, relative to blue. For a different star, we might measure just how much more blue light is coming to us, relative to red. From this kind of measurement we can determine the most commonly produced color of light for each star. A star that produces a lot of red light, like Betelgeuse, will have its most commonly produced color sitting very close to what we would call red, or perhaps slightly beyond what we can see, in the infrared.

▲ An artist's impression of exoplanet KELT-9b orbiting its host star, KELT-9. The KELT-9 star itself is only 300 million years old, which is young in star time—our own sun is 4.5 billion years old. KELT-9, by contrast, is more than twice as large, and nearly twice as hot, as our sun. It is unclear whether or not this planet can survive in its current orbit; this illustration shows its theorized eventual evaporation.

◄ The constellation of Orion is captured in this wide-field image, taken as a mosaic. Betelgeuse, the bright orange star in Orion's shoulder, is seen in the upper right. The constellation of Orion is also a gas-rich region of the sky.

A blue star, on the other hand, will have a most common color that is either blue or beyond blue, in the ultraviolet. And our white-light Sol? Sol's most common color is an eye-bending yellow-green. Rather than using the average color to discuss stars, it's usually this most frequent color that is used. This is how our star became known as a yellow star, even though the light is white to our eyes.

## COLOR AND HEAT

The color of a star isn't randomly assigned—it's a direct result of a few of its other properties. The color of each of the stellar siblings in our galaxy is determined directly by the temperature at the surface of the star. Temperature is in turn ruled by the mass of the star, along with how old it is.

The link between temperature and color is one that we encounter here on Earth as well, and it's evident in our language. We have turns of phrase for things that are "red-hot" and "white-hot," and we know that metal glowing orange is hotter than something glowing red, though both are dangerous to human skin. What we may be less familiar with is how far this relationship goes. Red-hot, for a star, is as cool as they come; the plasma that makes up the star is hot enough to glow of its own accord, but not hot enough to create a lot of blue light. Blue light needs more energy to produce, so at low temperatures, which means relatively low energies, there's just not enough energy to pump out a lot of blue light.

STARS

As the temperature of the star rises, the amount of available energy rises as well, and so more and more high-energy (blue) light can be produced. Stars will brighten from a deep red, to a brilliant white, to a vivid blue, if they are hot enough. The hottest stars produce a large amount of ultraviolet light, beyond the visible range, in addition to bathing their surroundings with blue light. These bright blue stars are relatively hostile to life; the moderate dose of UV light we receive from our Sun is already dangerous to humans, and luckily our atmosphere's ozone layer interferes with this light, blocking the higher energies from reaching the surface. With too much high-energy radiation, however, atmospheres can be destroyed, and life forms can be irradiated into oblivion. While these stars might have their own planetary children, hot blue stars are unlikely to have life-form grandchildren unless that life is particularly well sheltered or much more resilient against radiation than we are.

## MASSIVELY HOT

So color is the direct consequence of temperature, and temperature is, in turn, dictated by the mass of the star. For the set of our stellar family members that are, like our Sun, in a stable middle phase of their lifetimes, this is a fairly direct relationship. The more massive these stars are, the hotter their surface.

A star that contains a lot of mass is a stronger gravitational influence on the space surrounding it than a lower-mass star. This increased influence is also present on the material of the star itself. As we saw earlier, all stars are a balancing act between the inward, crushing pressure of gravity and the outward pressure of the stuff of the star resisting that inward force. With a stronger inward force of gravity due to the greater mass of the star, stability depends on an increased outward pressure. Without it, the star will collapse and become spatially smaller and denser.

> ## With too much high-energy radiation, however, atmospheres can be destroyed, and life forms can be irradiated into oblivion.

Any time you compress a gas (and all stars are made of gas), the temperature rises, and as the temperature rises, the outward pressure increases. If the star collapses down a little bit, the temperature inside the star increases, which increases the pressure pushing back against the force of gravity. As long as this happens slowly, the star can find a balancing point where inward gravity and outward pressure cancel each other out.

The spatial size of this end result also matters— a massive star ends up taking up more physical space once it has found its balancing point. If you have a larger glowing object, you illuminate the space surrounding your object more strongly. This tie between the size of a star and its brightness is so strong that if you doubled the size of a star, you would expect it to become four times as bright, if the temperature remained the same. There's a much stronger relationship between temperature and brightness: if you double the temperature, but keep the size the same, you'd expect the brightness to increase by 16 times. These relationships are so critical to understanding how stars function that the diagram most commonly used to describe a population of stars, the Hertzsprung-Russell diagram, or H-R diagram (shown opposite), plots temperature against their brightness. Both temperature and size change smoothly with each other, though, so you wouldn't expect to be able to keep one of them the same as you dial up the other.

▼ **Hertzsprung-Russell diagram**

The temperatures of stars (horizontal axis) are plotted against their luminosities (vertical axis). The position of a star in the diagram provides information about its stage of life and its mass. Stars that fuse hydrogen into helium lie on the diagonal branch, the so-called main sequence. Red dwarfs lie in the cool and faint corner. When a star exhausts all the hydrogen, it leaves the main sequence and becomes a red giant or a supergiant, depending on its mass. Stars with the mass of the Sun that have burned all their fuel evolve finally into white dwarfs.

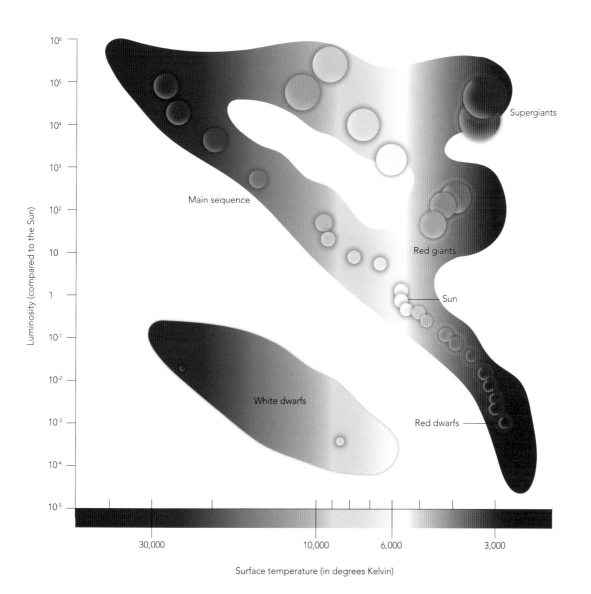

Luminosity (compared to the Sun)

$10^6$
$10^5$
$10^4$
$10^3$
$10^2$
$10$
$1$
$10^{-1}$
$10^{-2}$
$10^{-3}$
$10^{-4}$
$10^{-5}$

Supergiants

Main sequence

Red giants

Sun

White dwarfs

Red dwarfs

30,000    10,000    6,000    3,000

Surface temperature (in degrees Kelvin)

# HOW MANY TYPES OF STAR ARE THERE?

With the combination of the temperature of the star, as we trace it by its color, and the brightness of the star, which traces its physical size, we can piece together a lot of information about the vast family of stars that inhabit the Milky Way.

Stars that contain more mass than the Sun, and which are physically larger, have a much higher gravitational pull than our own star. In resistance to the greater inward force, the outward pressure has grown much higher, cranking the temperature up internally. The surface temperature of such a star mirrors that internal temperature, making it glow a hotter, bluer hue than our own Sun. Because the star is also physically larger, its brightness is also more intense than the Sun.

We also have Sol's smaller siblings, stars with much less mass than our star. These have less mass to crush inward, and so they find a balance with less outward pressure from the material within the star. This less-dense balance means that the core of the star is never pressed into extremely high temperatures. With a lower temperature in the core, the surface is even cooler, and only manages to glow a dim red. The smaller amount of material also means a smaller star, which drops the brightness, on top of it only glowing a duller red color.

These two scenarios, plus our Sun, give a range of stellar possibilities: from bright, large, and blue, to faint, small, and red. But is that the only set of stars you can get? Or are there exceptions to this? To answer that question we have to take a census of the nearby family of stars, and check what the rest of them are doing. If we find that the brightest stars are all blue, and the faintest of them are all red, then the description above is as far as we need to go.

A family census like this is best illustrated by charting the brightness of a large number of stars against their respective colors on the H-R diagram (see page 121). If our faint red to bright blue relationship describes all stars, then we should expect to see a diagonal line running across the diagram, with nothing anywhere else.

> In resistance to the greater inward force, the outward pressure has grown much higher, cranking the temperature up internally.

## OUR OWN YELLOW DWARF

Our Sun, which sits roughly in the middle of the H-R diagram, is just on the dwarf side of the giant/dwarf divide; technically speaking, our star is a yellow dwarf. We've talked extensively about how our own star works, and its twin stars, other yellow dwarfs (they bear tangled names, from Tau Ceti to HD 147513), so far seem to be operating in the same way.

▶ The star V838 Monocerotis, as viewed by Hubble on December 17, 2002. This star produced an unusual outburst in January 2002, and the surrounding cloud of dust and gas is still being illuminated by the burst of light.

We do see the diagonal line. This line is what we call the "main sequence" of stars, and it is where all of the stars that are stably burning hydrogen in their cores sit. The vast majority of the stellar family members in our neighborhood and galaxy sit along that line. However, our Sun has a few unusual siblings, which are found in a totally different part of this diagram.

It's useful to categorize stars more carefully than just into red, blue, and yellow like our Sun. The H-R diagram is a good place to start dividing the stellar family into sets of cousins, where the stars within that group are all pretty similar to each other and can be compared as a group to other groups.

The first broad division we can make is a horizontal one, across the middle of the H-R diagram. The top half of the diagram is where all the extremely bright, physically large stars in our galaxy can be placed. The bottom half gives us the smaller, dimmer stars. These smaller stars are formally given the term dwarf stars, distinguished from their larger cousins, called giants. This giant/dwarf divide is particularly useful for stars that glow an intense red, as a red star can either be a small star along the main sequence, or one of the giant stars, well away from this main sequence line. If we use the color of the star, and whether it is a giant or a dwarf, we can easily describe any star in our home galaxy—and beyond.

# WHAT ARE DWARF STARS?

In the previous section we looked at a few types of stars, but what of the others? There's more to the stellar population than just the large stars that form the brightest objects in the sky. Let's begin with the other dwarfs. Once we exit the yellow dwarf area, we have a few stellar colors that remain: red dwarfs, brown dwarfs, and the only small stellar sibling of ours that is bluer—white dwarfs.

### RED DWARFS

Red dwarfs are one of the closest relatives of our Sun—just a little less massive than Sol, they also sit along the main sequence, toward the fainter, redder end. Red dwarfs, much like the Sun, burn hydrogen at their cores, but have gathered much less material to themselves in their formation. This smaller amount of mass means that the pressures and temperatures at their cores are lower, and they burn at a slower rate as a result. The slower burn means that these dim, redder stars will outlive their bluer siblings. Their smaller mass means that they're relatively easy to form, and are expected to be among the most numerous type of star in our stellar family. It's hard to count them directly, because they are so faint and red, but we know that our nearest stellar neighbor, Proxima Centauri, is a red dwarf star.

If you keep pushing down the main sequence to even fainter and dimmer stars, you arrive at an interesting corner of our stellar family. By definition, stars are objects that can construct hydrogen into more elaborate elements through nuclear fusion, and this process operates more efficiently the hotter and denser the core of your star happens to be.

### BROWN DWARFS

At some point, the core of the star just isn't hot and dense enough to build anything. Brown dwarfs are, in essence, failed stars. They collapsed, gathered material to themselves, and began to heat up, but there just wasn't enough material around to build a fully fledged star, and so they were not able to build up the elemental complexity of our universe.

> By definition, stars are objects that can construct hydrogen into more elaborate elements through nuclear fusion, and this process operates more efficiently the hotter and denser the core of your star happens to be.

Unable to generate their own heat, brown dwarfs have to be content with simply radiating out into the cosmos what heat they generated during their collapse, slowly cooling to the average temperature of the universe. This process takes an extraordinarily long time, so we don't expect any of them have succeeded in getting rid of all their heat—but if they had, we'd call them our Sun's black dwarf siblings.

Brown dwarfs are essentially very large Jupiters, though brown dwarfs are counted as stars while their Jupiterlike counterparts are always planets. They can be "cold" even on human scales: a brown dwarf found in 2011 was only as warm as a cup of coffee. Their masses are so much smaller than our own star that instead of being discussed in units of multiples or fractions of the mass of the Sun, we use the mass of Jupiter instead. If the star is more than 83 times the mass of Jupiter (or 0.08 times the mass of the Sun), it can start to burn hydrogen in its core. Brown dwarfs then range from this upper limit down to somewhere around 13 times the size of Jupiter.

There's nothing special about this 13 Jupiter mass cut-off for how small a brown dwarf can get. At that size, it's debatable whether you should call this collection of gas a very large planet (which would usually orbit a fully-fledged star), or a very small star (which would usually be at the center of its solar system). If the brown dwarf/large Jupiter happens to be orbiting a much larger star that is burning hydrogen in its core, it's even more ambiguous. Binary stars are common in our galaxy—like twins, being born out of the same cloud of gas, so it's not out of the question to have a brown dwarf twinned with a yellow dwarf like our Sun. But when the brown dwarf is particularly small, like a large planet, orbits like a large planet and isn't generating its own heat (like a planet), it's easy to begin to classify these objects as planets instead of stars.

▲ This is an artist's concept of the red dwarf star CHRX 73 (upper left) and its companion CHRX 73 B in the foreground (lower right) weighing in at 12 Jupiter masses. CHRX 73 B is one of the smallest companion objects ever seen around a normal star beyond our Sun; in terms of classification, it resides almost exactly on the boundary between a binary star system and a brown dwarf with an exceptionally large planet.

▲ An artist's conception of the brown dwarf 2MASSJ22282889-431026. Brown dwarfs' atmospheres can be similar to the giant planet Jupiter's. The Spitzer and Hubble space telescopes simultaneously observed the object as it rotated every 100 minutes, finding that there are likely wind-driven, planet-size clouds.

The smallest brown dwarfs, if they should be considered planets, would be the children of the stars; but if they are the smallest stars, then they must be placed one tier up, as the potential parents to their own planetary children. This kind of bridge across the generational gap between stars and planets is an unusual one, and it's part of what makes brown dwarfs so intriguing. The largest planets in our solar system (Jupiter, Saturn, Uranus, and Neptune) are typically dubbed "gas giants," and it's in the gap between Jupiterlike planets and the stars that illuminate those planets that brown dwarfs find themselves.

# WHITE DWARFS

Aside from the red giants, white dwarfs make up the other population of stars that is very far from the main sequence—in fact white dwarfs are the only kind of star that falls below the main sequence. This is a weird place for a star to be, because it means that the star is, for some reason, burning relatively hot, but is also extremely faint. If the star is hot, the only way for it to be so faint is for it to be very, very small. As we've seen, small stars usually don't have a lot of mass, which is how you get faint red stars—so why are these burning so hot?

White dwarfs, unlike our Sun, are not burning hydrogen in their cores. In fact, they're the leftover core of a star, hyper-compressed into an extra dense state. White dwarfs are often about the physical size of Earth, but they contain large fractions of the mass of the Sun. This ultra-high density means that the material within the star exists in one of the most extreme places in the universe. It also gives us the explanation for the location on the H-R diagram. The white dwarf siblings of our Sun are tiny, which explains their dimness in our universe. However, their huge amount of mass accounts for their white-hot glow, given the intense inward crushing force that gravity is exerting on the star.

A normal star has an outward pressure to balance gravity, as we have seen, and white dwarfs must have this as well, or they would simply continue to get smaller as time progresses. We see them as stable siblings of our Sun, so what generates this resistant force? The mere presence of electrons is what's doing it. At the incredible densities of a white dwarf, electrons resist further gravitational crushing. If you wanted to compress a white dwarf farther down, you'd need to compress electrons closer in to each other. The typical white dwarf doesn't have the gravitational force required to mash electrons any closer together, and so they can put up an effective resistance, holding the star from becoming any smaller.

▲ Cutaway illustration of a white dwarf, the end state of stars around the same mass as our Sun. Some white dwarfs may contain crystallized solid cores.

▼ **Mass–radius relation for white dwarfs**
As the mass of a white dwarf increases, the electrons of its atoms are compressed more closely together, which causes its radius to decrease. Electrons produce a finite resistance to gravity; as a white dwarf's mass increases, the inward pressure increases, and the stellar remnant shrinks. This process fails at 1.4 stellar masses (known as the Chandrasekhar limit), at which point a white dwarf becomes unstable and detonates.

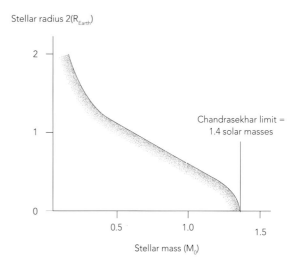

Stellar radius 2($R_{Earth}$)

Chandrasekhar limit = 1.4 solar masses

2

1

0

0.5    1.0    1.5

Stellar mass ($M_0$)

STARS

# What would happen if we split the Sun in half?

There's no actual force in our universe that could be responsible for taking one stable star and dividing it, like dough, into two stars half as big, but that's what thought experiments are for. What would happen to Earth, assuming it stayed orbiting where it is, around our newly formed double-star system?

Fundamentally, we need to pose a couple of questions: How does the amount of energy produced by a star change as we alter its mass? And what effect would any change have on Earth? If the amount of light and heat has a linear relationship to mass, two stars of half the mass of the Sun should sum up to effectively the same thing as one Sun-sized star. But if that's not true, we could wind up with more sunlight (baking Earth) or much less (freezing Earth).

### 1. How much sunlight?

Stars of approximately half the mass of the Sun are classed as red dwarf stars. Let's do ourselves a favor and assume that these two stars orbit each other closely and stably, so we don't have to worry about their orbits changing over time. A double-red dwarf system, since its total mass is the same as the Sun, would be gravitationally similar to our own Sun, and so none of the orbits of the planets would have to change.

However, as we've touched on before, the relationship between the amount of light that's produced and the mass of the star is not a one-to-one trade-off. If I cut a star in half, I halve its mass, and I drop the amount of light produced by that star by 90 per cent. The drop in temperature between Sol and a red dwarf is also significant: about 40 per cent less. Our Sun's surface temperature is 9,980°F (5,527°C), whereas a red dwarf of half the mass only checks in at 6,200°F

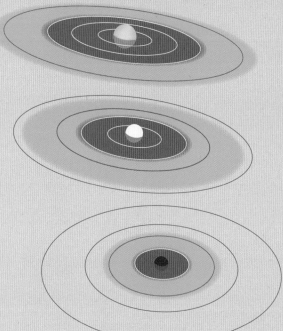

◄ **Where liquid water could occur**

The orbits of the inner rocky planets in our solar system here indicate scale, with regions around stars like our Sun shown in the middle. In blue is the zone in which liquid water is possible. The outer edge indicates where water would freeze, and in red where it should boil. A cooler star (bottom) would have liquid water possible at Venus' orbit, while a hotter star would make liquid water stable at the orbit of Mars.

(3,426°C). This drop in temperature is also responsible for a dramatic shift away from white light toward the red end of the visible spectrum.

## 2. How would Earth be affected?

This double-star system would only be able to produce 20 per cent of the light that the Sun generates. That is a pretty dramatic reduction in the brightness of the center of our solar system, which would, in turn, dramatically shift the places surrounding that double-star system where liquid water could exist. A planet like Earth, orbiting that double-red dwarf system at a distance of 93 million miles (149.5 million kilometers), would not be able to keep water from freezing. As far as we can tell, liquid water is the primary requirement for life to exist, so an Earthlike planet around such stellar parental twins would be frozen solid.

As we saw in Chapter 3, the liquid water zone is calculated as the shell of space surrounding a star where there's enough energy reaching the surface of the planet for water to melt, but not so much that the water would evaporate away as steam and be lost to space. The formal name for this region of space is the habitable zone, and it's the focus for many searches for plausible Earthlike cousins outside our solar system.

### Verdict

In a double-red dwarf star system, the habitable zone would be so close to the stars that if we placed our own planets around those twin stars, Mercury would be in the liquid water zone. Mercury's orbit sits between 0.3 and 0.46 times the distance between Earth and the Sun. The habitable zone would range between almost these same numbers: 0.32 au, if the planet was getting light from only one star, up to 0.44 au, if it was getting sunlight from both of them. So Earth would be completely frozen, but the water ice trapped in Mercury's darkest craters would be able to thaw, forming crater lakes—and given enough time, possibly life.

## CHAOTICALLY CREATING ROGUE WORLDS

Why does it throw planets out of a solar system if you have two stars orbiting in changing orbits? This is because you start to run into what's formally called "three-body interactions," and these are fundamentally chaotic systems. If two objects in the three-body system pass close enough together, they can exchange energy. If the smallest object donates energy to the more massive one, nothing much happens, but if the larger one donates energy to the smaller one, that energy donation can be enough of a kick to its velocity that it might go veering out of the solar system entirely, wandering the galaxy as a "rogue planet." Binary systems have been found to have planets around them (next time you see a news article claiming to have found a Tatooine, that's what this means), but in general the planets are *very* far away from their stars, to the point where the planet doesn't care if it's orbiting one star or two, and the stars themselves are typically in wider, stable orbits. Triple star systems can be stable, but they often settle into a close binary and a distant third object. If you have three objects orbiting close to each other, this will nearly always be fundamentally unstable; one of the three will wind up being flung out eventually, though "eventually" is on an astronomical time scale and might be a while.

# HOW ARE THE STARS GIVEN THEIR NAMES?

It's one thing to be able to describe our stellar family in groups of similarly behaved stars, and another thing entirely to give them each a name. As much as we call our own star "the Sun," its formal designation is "Sol." We might know a few other names of Sol's siblings—the North Star, for instance, is named Polaris, and Betelgeuse, a red supergiant star, sits at the constellation Orion's shoulder, as we noted earlier.

But what about the other stars? If you continue digging for names, you'll start to find ones that are less familiar to your tongue, unless you happen to speak Arabic. We don't have to leave the constellation of Orion to find a few: the stars in Orion's belt are named Alnitak, Alnilam, and Mintaka. These names are old; as long ago as 150 CE, in the era of the mathematician and astronomer Ptolemy, people had begun to collect the names of stars into catalogs. Ptolemy himself developed a catalog that was translated from Greek into Arabic, and that Arabic catalog was then circulated around western Europe.

Many of the Arabic names are simply descriptions of the location of the star within the constellation—stating that they are in the belt, for instance. In the process of coming from the Arabic translation back into western Europe, many of the spellings were altered, and so some of these oldest names either have multiple spellings, or have converged to a spelling that has only a loose resemblance to the original Arabic. Mintaka and Alnitak, for instance, could have been more accurately written out as al-Mantaqa and an-Nitāq. But these are only the brightest of the night-sky stars, and of course our telescopes have improved immensely since the days of Ptolemy, vastly growing the list of our known stellar family.

An attempt to make a comprehensive list of all the stars in the night sky was undertaken in the 1600s by Johann Bayer, and this has since taken on his name as the Bayer Designation system. The idea was to adopt a more rigorous approach, where for every constellation, each star could be named in a more catalog-friendly way, with a Greek letter to mark its brightness and a name to mark its constellation. The brightest stars in Orion then became Alpha Orionis, Beta Orionis, and so on, and fainter stars were assigned Greek letters farther down the alphabet. However, this alphabetical assignment is not always exactly in line with the brightness of the stars. Stars that vary in brightness can be particularly troublesome if you are trying to sort them in order.

The Bayer Designation was used for a number of years, but we have begun to take a huge number of photographs of the night sky, at ever fainter depths, and any attempt to maintain an all-sky catalog of stars with a consistent naming scheme has essentially been abandoned. Most of the stars that we identify in the night sky are now "named" with an alphanumeric identifier, which marks the exact survey of the sky that discovered them, and often their location. These stars will not gain more pronounceable names unless something very unusual about them is later discovered, and the scientists investigating that abnormality need a faster way of referring to the star.

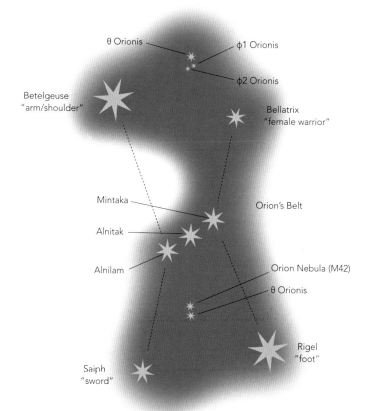

The stars of Orion's belt in a deep image of the night sky. Alnilam (just below and to the right of the yellow-orange nebula on the left), Alnitak (the bright star in the center), and Mintaka (top right) are shown, alongside the Horsehead nebula in the center left, and the Flame nebula (top left).

**◄ The constellation of Orion**

A star map of the constellation Orion, as they are more typically rendered in diagram form. Typically only the brightest subset of stars are shown.

θ Orionis

φ1 Orionis

φ2 Orionis

Betelgeuse "arm/shoulder"

Bellatrix "female warrior"

Mintaka

Orion's Belt

Alnitak

Alnilam

Orion Nebula (M42)

θ Orionis

Rigel "foot"

Saiph "sword"

# WHAT EXACTLY ARE CONSTELLATIONS?

Even though we've spent many hundreds of years basing our stellar family's names on them, the constellations themselves are nothing more than the storytelling human brain at work on the night sky. With a couple of exceptions, the stars in the constellations are no more related to each other than the lights of a plane flying at night are to the stars.

The stars in our constellations are usually fairly bright, which means that they can't be on opposite sides of the galaxy from each other, and they remain related by the simple fact that they are all in the same galaxy. But not being on opposite sides of the galaxy and being really and truly lined up in the night sky are two very different things.

## A SINGULAR VIEW FROM EARTH

The constellations are simply temporary alignments of the stars in our neighborhood, and they only exist as we see them from our particular vantage point in the galaxy. If our solar system were even a little bit differently placed in the galaxy, those constellations would transform into a completely different jumble of stars relatively rapidly. Orion is a good example of this. It's a bright constellation, with a number of the brighter stars in the night sky of the northern hemisphere contained within it. In winter it's easily recognized, even in cities where the light pollution can be severe. And yet its stars are hundreds of light-years apart from each other.

Betelgeuse, the bright red star in Orion's shoulder, is about 640 light-years from Earth. In contrast, Bellatrix, the other shoulder star, is 200 light-years from Earth, three times closer. Mintaka, the star farthest to the right in the belt, is a whopping 1,200 light-years away, twice as far as Betelgeuse. If you put those distances into a model of the constellation, and rotate around it with an arbitrary camera, you can see how dramatically the constellation changes as your perspective changes.

This isn't unique to Orion. Effectively any constellation you could point to would do the same thing as we moved around it. The slightest shift in where our solar system is placed in the galaxy would scramble the entire constellation. We're not ever going to move the solar system, but there's another way for the constellations to scramble themselves: waiting for a long, long time.

## THE BIG DIPPER

The Big Dipper (also known as the Plow) is also made up of stars that are unrelated to each other, but those stars are all much closer to us than the stars in Orion. The seven brightest stars that make up the Big Dipper range from 58 light-years to 124 light-years distant from Earth. Much in the same way as trees by the side of the road appear to zip past your eyes faster than objects in the distance, stars that are closer to our position appear to move much faster in the skies than distant stars. So the stars in the Big Dipper will appear to move over time more rapidly than those in Orion. We can measure the speeds at which these stellar family members are moving, and predict where they will be in the future. In another 100,000 years, the handle on the Big Dipper will be bent under, and the front of the Big Dipper bent outward. It's a relatively quick destruction of an astronomically temporary alignment.

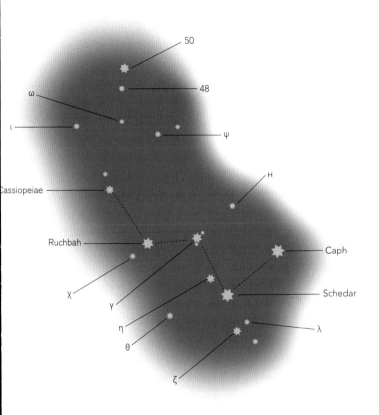

50
48
ω
ι
ψ
H
Cassiopeiae
Ruchbah
Caph
χ
γ
Schedar
η
θ
λ
ζ

▲ This sky map shows the "W" shape of the constellation Cassiopeia. Nearby, the location of the star HD 219134 (circled), is marked, itself host to the nearest confirmed rocky planet found to date outside of our solar system. The star can be seen with the naked eye in dark skies. It actually has multiple planets, none of which are habitable.

◄ **The constellation of Cassiopeia**
A map of the stars in Cassiopeia. Cassiopeia is a circumpolar (meaning its stars never rise or set, but circle the pole star) constellation in much of the Northern Hemisphere.

# WHAT ARE GLOBULAR CLUSTERS?

Many of the stars in the night sky are solitary, completing their orbits around the center of our galaxy on their own, or perhaps with a few planetary children to keep them company. Our Sun is one such star—we have no secondary sun in our skies, and the Sun orbits the center of the galaxy with the solar system coming along with it. But not all stars are so isolated, and there's another assembly of stars that's worth a look before we explore all the varied ways that stars can come to the ends of their lifetimes.

Globular clusters are very dense, very old clusters of stars, where all the stars are gravitationally bound to each other, so that it will be very difficult for them to ever escape. These are a family group; gravitationally tied together, and usually made up of stars that are all the same age, it's thought that they formed nearly all at once and have remained clustered together as a tight-knit community. These clusters are often found outside the main disk of the galaxy, in a spherical arrangement instead of the flat, disklike arrangement of the rest of the stars in the Milky Way. They can be quite isolated from the rest of the stellar family within the galaxy, and they contain a much older population of stars that are many billions of years older than our own Sun.

Globular clusters are home to a few hundred thousand stars on average, and our Milky Way hosts a few hundred of them. (The galaxy Andromeda, our closest galactic point of comparison, has at least 60 globular clusters, though we're bound to miss some when we're hunting for a faint collection of stars 2.5 million light-years away.) Sometime in the formation of the Milky Way, these clusters of stars were formed and trapped into spherical groups around the galaxy.

It must have been early in the formation of the galaxy, because the stars that populate globular clusters are very, very old. We mentioned earlier that small, red stars take much longer to exhaust their supply of hydrogen gas, and can stick around as stars for a very long time. Some of the stars in globular clusters appear to date back nearly 13 billion years—about as old as you can get. In fact, for a time, the age of these stars was used to age the universe as a whole. If you've got stars hanging around that are 13 billion years old, the universe had to be older than that. But their exact formation is still a mystery, and how they manage to remain unperturbed for such a long stretch of time, while the universe around them has changed so drastically, is another point of study.

## Sometime in the formation of the Milky Way, these clusters of stars were formed and trapped into spherical groups around the galaxy.

▶ The global cluster NGC 1466, viewed here through the NASA/ESA Hubble Space Telescope. It is a gathering of stars held together by gravity that is slowly moving through space on the outskirts of the Large Magellanic Cloud, one of our closest galactic neighbors. This globular cluster is roughly 13 billion years old.

## THE SECRET OF LONGEVITY

One thing is for sure: if the globular cluster was going to collapse onto itself, or flatten itself, or change in any way, it's had more than enough time to do so. Since we're seeing them now as they are, they must be stable in their current setup. So what keeps all these densely crowded stars from collapsing inward on themselves? We've talked a lot so far about the influence of gravity on individual stars. If you crowd 100,000 stars into a fairly small region of space, you might very reasonably expect that gravity should pull them all down into each other, collapsing the whole lot into a messy superstar at the very center of what was once a globular cluster. If gravity were the only force at work here, that's exactly what would happen.

But gravity is almost never the only force at work, and globular clusters are no exception. In this case, the missing piece of the puzzle is that every single one of the 100,000 or more stars in the globular cluster is in motion. And with motion, we can defeat gravity in the same way that the International Space Station stays aloft: by orbiting. If you have enough motion "sideways" from the direction that gravity is trying to pull you, the end result is an oval orbit around the center of mass. For the ISS, this is a very nearly circular orbit around our parent planet. For a star in a globular cluster, its oval pathway takes it in loops around the center of the cluster. As long as that sideways motion persists, the star won't fall to the center of the group.

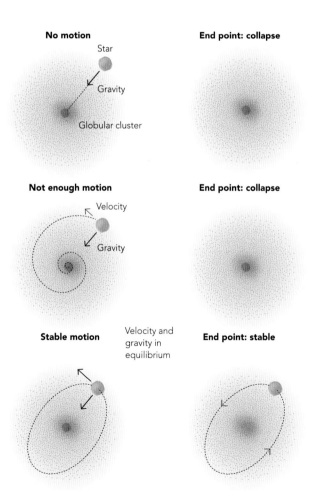

**No motion**

Star

Gravity

Globular cluster

**End point: collapse**

**Not enough motion**

Velocity

Gravity

**End point: collapse**

**Stable motion**

Velocity and gravity in equilibrium

**End point: stable**

◀ **The dynamical options for a star's motion**

If a star is held "stationary" relative to a massive object, gravity will eventually pull it to the center. If a star has a little bit of sideways motion, the force of gravity is unchanged, and the star will collapse down into the center on a longer, spiraling path. If the star has enough velocity in a sideways direction, the motion can allow the star to continuously fall in an elliptical path around the central object, maintaining a stable orbit.

▲ The sharpest image ever taken of the large "grand design" spiral galaxy M81, seen by the Hubble Space Telescope. While the disk of the galaxy is dominated by individual stars, and small groups of stars, the globular clusters that surround this galaxy are easiest to see far from the center, where they inhabit a sphere of space surrounding the entire galaxy.

## TRAILING STARS

There's one other thing that can cause a lot of change within a globular cluster. The clusters are all in orbit around a parent galaxy—and sometimes they wander too close. The orbits of globular clusters around their galaxy are extremely long, and many of them seem to have survived many such orbits, but it's not always a safe passage.

If the cluster travels too close to the galaxy, the galaxy's gravitational force can shear apart its outer layers. If the cluster has already sorted itself, that means it loses the lightest, smallest stars from its outer edge. These are pulled into a faint stream, barely visible even to powerful telescopes, and leaving the high-mass, larger stars at the nucleus of the cluster. At least four such globulars exist within our own Milky Way.

Every single one of the stellar cousins in the globular cluster is using the exact same strategy, with absolutely no rhyme nor reason to the direction of "sideways" they choose to go. The randomness of the stars' orbits is what makes the cluster look so spherical. If the stars all had the same definition of "sideways," like the planets in our solar system do, you'd wind up with a disk, and the cluster would look flattened.

This very specific kind of structure—a random set of orbits, from a large number of objects, creating a fuzzy, spherical haze of stars, is given the technical name of a "pressure-supported" structure. There's no real pressure here, like there is on the inside of each of the stars, but it serves as a distinction from the ordered rotation of a solar system disk or galaxy disk, and it is still a resistance to the force of gravity. (Astronomers are bad at naming things.)

It is the motion of the stars that lets these globular clusters remain so stable over the lifetime of the universe. But globular clusters aren't totally immune to changes over the course of 13 billion years.

## OCCASIONAL INTERACTION

The stars within a globular cluster are more tightly packed by far than the stars within the disk of the galaxy—typically they wander within a light-year of each other, though at the very center, where the stars are the densest, they can come as close as the distance between the Sun and Pluto (around 3.7 billion miles, or 5.95 billion kilometers, which is not very close, but in astronomical terms it's snug.) As densely packed as they are, the stars do occasionally interact with each other when they get close enough; as they swing by each other, the gravitational force will tug on them. If the two stars are about the same size and mass, and they're moving in the same general direction, but one has a much faster orbit around the center of the cluster, it may overtake the slower star. The gravitational tug between the two will slow down the faster star, donate that energy to the slower

star, and speed it up. Because they have similar mass, the two stars can end up traveling much closer to the same speed after their encounter.

But what happens if the second star is much less massive? In that case, the same amount of energy donated to the small star can bounce that smaller star to a much higher speed. The change in speed of any star is related to the energy donated, and its mass. The less energy, the less of a boost; and the less massive the star is, the more of a boost it gets. The momentum of the smaller star is less able to resist changes in speed, and off it zooms. This is the same principle behind sling-shotting spacecraft around Jupiter; we steal a small amount of energy from Jupiter, but because the craft is so little, it can increase its speed dramatically.

## STELLAR SEGREGATION

If you let this energy donation process proceed for long enough, what happens is that the massive stars will always donate energy to less massive stars and will gradually slow down. Meanwhile, the less massive stars, which are on the receiving end of all these donations, gradually speed up. These speeds translate into changes in their respective orbits. If the star has slowed down, gravity has an upper hand, and so it pulls the star down into the depths of the cluster. If the star has sped up, it will end up flinging itself around the outskirts of the cluster at high speeds, in sharp contrast to the now lethargic orbits of the largest stars. The globular cluster has effectively sorted itself, with all the heaviest stars at its center, and all the lightest, smallest stars out at the edges—a process called "mass segregation."

## ▼ The distribution of stars in a globular cluster

1. If you build up enough elliptical orbits and then capture stars at random locations, you will rapidly build up a circular-ish blob of stars, more diffuse at the edges, and more concentrated at the center. Each star is (roughly) independent of all the others, and traces out its own unique path through the globular cluster.

2. Mass segregation is the process by which heavier stars settle at the center of a globular cluster. The larger mass stars, represented here as the large dots, will in time find themselves preferentially in the center, while the low mass stars (shown here as small dots) wind up at the outskirts. If the globular cluster finds itself a bit too close to a galaxy, these outer layers will be lost to the gravitational influence of the galaxy first.

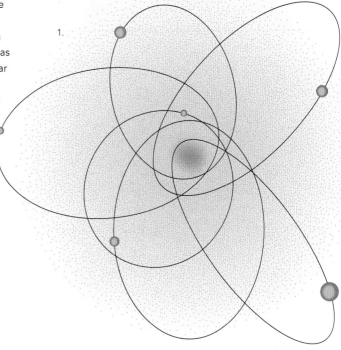

1.

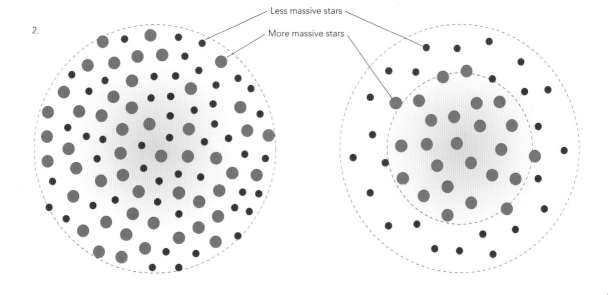

2.

Less massive stars

More massive stars

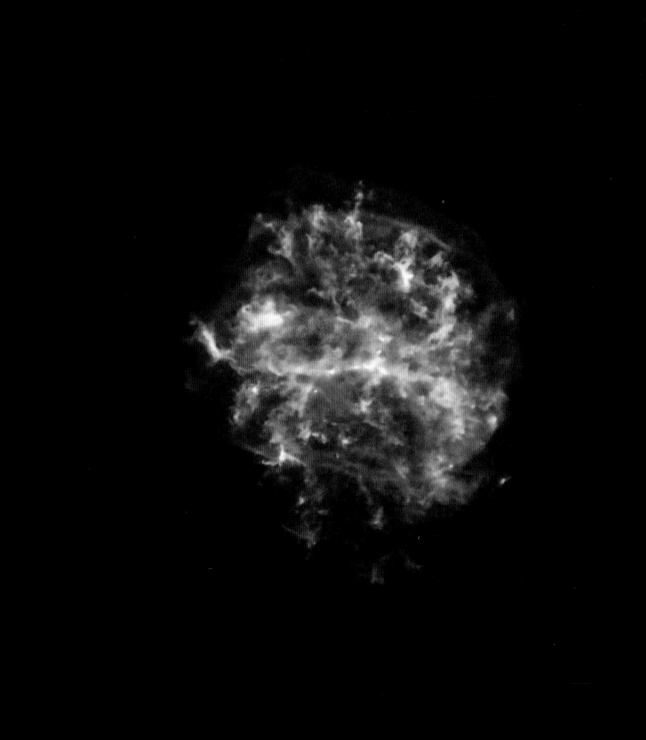

# STELLAR DEATHS

In the history of the universe, many generations of stars have formed, burned, and died, scattering material back to their home galaxies. But not all stars act the same when they reach the ends of their lives, so let's look at the dramatic finales of stars' lives.

# HOW DO STARS DIE?

The lifetimes of the stars dwarf our human ones: from the first fusion in its core to the end, our Sun will have provided around 8 billion years of relative stability. But all stars have a fixed lifespan, tied closely to their mass and the rate at which they run through the hydrogen available to them.

The main sequence of stars, that diagonal line from bright and blue to faint and red (see page 121), is composed entirely of the siblings of our Sun that are stable and burning hydrogen into heavier elements in their cores. These stars maintain an outward pressure, resisting the gravitational force pressing inward because they are burning so much hydrogen, and keeping the internal temperature high. As time goes on, more and more of the available hydrogen is consumed and turned primarily into helium. At some stage there isn't enough hydrogen around to continue burning.

## SUCCUMBING TO GRAVITY

A dwindling supply of hydrogen marks the beginning of the end for a star. It must, at this point, undergo change. Without the ability to continue burning hydrogen into helium in its core, a large part of its resistance to gravity is gone. We should expect the star to compress itself as gravity wins out. This is indeed what happens; the star's interior is crushed to even higher densities, pressing the plasma into a hotter, more confined space. This crushing will continue until a region surrounding the core of the star—now filled with helium—can begin to burn hydrogen. The core may stay inert, because as far as the hydrogen-burning process is concerned, helium is the ash left over after a camp fire. If the star is large enough, the pressures in the center may increase until the core begins a brand-new fusion process, building helium into elements like carbon and oxygen.

Our grandparent Sol will be able to burn helium like this, but the more massive siblings of the Sun will be able to push even farther. Once the helium in the core of the star is exhausted, the star will collapse again, triggering hydrogen burning in a new shell of material even farther from the core, plus helium burning in the shell where hydrogen had been burning, and the carbon and oxygen ash from the helium burning may ignite. This process of exhausting the fuel at the core, allowing the star to collapse farther, and igniting another layer, allows extremely massive stars (more than eight times the mass of the Sun) to build their way up to iron ash in their cores. That is the end of the line—the pressures and temperatures required to build the heavier elements cannot be produced through gravitational pressure alone.

Even with just two shells of burning, as our Sun will do, there's a lot of extra outward pressure from within the star. A large amount of the mass of the star is in its depths, where the material is compressed. At the surface of the star, the atmosphere is much thinner and more tenuous, and thus contains less mass. With a boost to the amount of light being produced in the depths of the star, this outer layer is hit with an extra strong outward push. Not being very massive, and being relatively far from the majority of the mass of the star, these outer layers are forced outward.

Several things are now happening at once. As the star's outer layers expand, they cool. It's the opposite effect that's happening in the core, which is heating up as it compresses. This takes the surface of the star from a white or blue color down to an orange or a red, as its surface temperature drops. At the same time, the interior of the star is burning through helium and hydrogen, both at a furious pace, and the

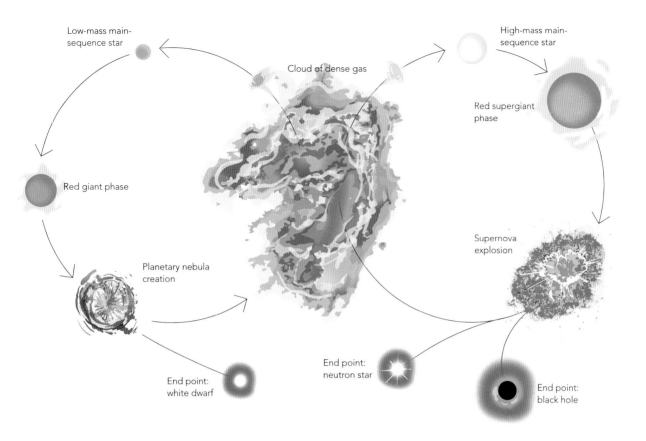

Low-mass main-sequence star

Cloud of dense gas

High-mass main-sequence star

Red supergiant phase

Red giant phase

Planetary nebula creation

Supernova explosion

End point: white dwarf

End point: neutron star

End point: black hole

amount of light being produced has ramped up. Even after cooling down its surface, the star ends up thousands of times brighter than it was back when it was only burning hydrogen.

With all these changes going on, the star is going to move on the H-R diagram (see page 121). No longer behaving like its main-sequence siblings, the end-of-life star moves to the right of the diagram as it cools to a redder color, and up, to the brighter end, as its light-producing reactions grow more rapid. We've created a red giant star. All stars, except the very smallest red dwarfs, will pass through this phase on their way out of stellar life. Our own star should do so in the next 4.5 billion years or so.

### ▲ The life cycle of stars

Every star begins in a star-forming gas cloud. Whether that cloud is large enough to collapse into a massive star, or into one more like our Sun, will dictate its future path. The Sunlike stars will evolve along the pathway to the left; after billions of years they will form into a red giant, then a planetary nebula, and finally to a white dwarf. Massive stars, by contrast, only spend millions of years in existence before cycling through the right-hand pathway: a red giant and then a supernova.

# WHAT WILL HAPPEN TO EARTH WHEN THE SUN DIES?

The expansion of a star into a red giant is no small thing—when the Sun reaches this phase of its life, its surface will extend 200 times farther than it currently does, roughly to the orbit of Earth. Earth is not expected to survive the death of its parent star; even if it wanders a bit farther from the center of the solar system, most predictions do not expect our planet to totally escape the atmosphere of the expanding Sun.

If Earth is caught in the Sun's outer layers, it will be vaporized, and Venus and Mercury, our planetary neighbors, have no chance of escaping disintegration. It won't be an abrupt end for our planet—Earth will become gradually more and more inhospitable as billions of years run on. Even before the Sun finishes burning hydrogen, it will be slowly evolving. Since our star began building helium out of hydrogen in its core, it has increased its brightness by about 10 per cent every billion years. Increased brightness means the amount of heat our planet receives is also greater. With enough heat, the water on the surface of our planet will begin to evaporate.

A 10 per cent increase in brightness is enough to change the location of the habitable zone (see page 84) around our star, placing it outside Earth's orbit. As soon as Earth exits the habitable zone, we no longer have a planet that can stably store liquid water on its surface, and the evaporation of the oceans will begin in earnest. By the time the Sun stops burning hydrogen in its core and becomes a fully fledged red giant, Mars will be in the habitable zone, and Earth will be much too hot to maintain water on its surface.

As the oceans evaporate, the water will instead be held in the atmosphere. Water is an excellent greenhouse gas, much like carbon dioxide, and will serve to trap even more heat around the planet.

With the heat trap in place, more and more water will evaporate to join the atmospheric vapor, increasing the greenhouse effect until the oceans are dry. The atmosphere will be a muggy, superheated, water-saturated mess. The water molecules that have made their way to the upper atmosphere will get bombarded with high-energy particles from the Sun, breaking them apart. Once the hydrogen and oxygen that make up the water molecules are freed, they can escape from the atmosphere into space. Over time, this bombardment of the atmosphere by the Sun will bleed the atmosphere dry of water.

## COUNTDOWN

The trigger point, when the amount of light the Sun sends to the surface of our planet becomes too great for stable water, is predicted to arrive in about 1 billion years. The exact numbers depend on who you talk to; predictions differ in terms of how fast this whole process unspools. Our planet will certainly walk this path, but some scientists suggest that Earth will be roasted well before the 1 billion-year mark, because the rocks, oceans, and plate tectonics may all speed the drying and heating of the planet. On the other hand, some forms of life have already proven themselves to be extremely durable (see pages 64–65) and able to maintain a foothold in odd corners of Earth, living off of chemicals, like the deep sea creatures of today.

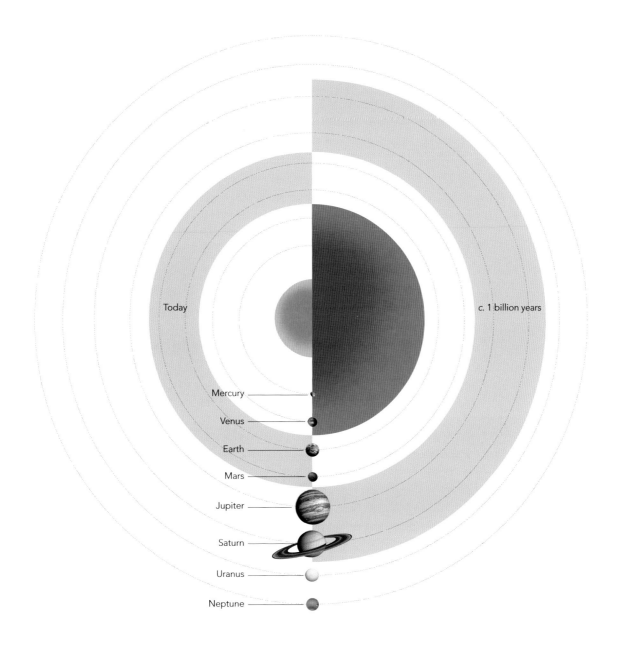

Today

c. 1 billion years

Mercury

Venus

Earth

Mars

Jupiter

Saturn

Uranus

Neptune

One billion years, or thereabouts, is as good a baseline as any in terms of the time remaining for life on our planet.

Yet however much damage and destruction our evolving star will wreak on the neighborhood, our Sun is still only a medium- to small-sized star, and so its ballooning out into the solar system is still relatively small-scale destruction.

▲ **Present and future habitable zones**
Currently, the habitable zone of the Sun encircles Earth's orbit. As the Sun expands into a red giant, the zone where liquid water is possible will move outward to where Jupiter's orbit is now (orbits not drawn to scale). The orbits of the planets will change as a result of the evolution of the Sun, drifting outward from their current locations.

# HOW BIG ARE THE SUPERGIANTS?

UY Scuti is one of the largest stars we know of. It's found in the constellation Scutum, near Sagittarius, along the line of the Milky Way in our night sky. UY Scuti isn't the most massive star—its gravitational clout is outdone by many others—but it appears to be taking up the most space of any star in our repertoire. UY Scuti is not a lightweight: it's a good 30 times more massive than our own star. However, because its outermost layers are already being nudged outward, the material that makes up the star already covers a region 1,700 times larger than the Sun.

**B**eing nearly 2,000 times bigger than our own Sun places UY Scuti into a new class of star—a supergiant. Betelgeuse, the red star in Orion's shoulder, is one of the most obvious supergiant stars in the night sky. Most supergiants are about half as large as UY Scuti, but they are nonetheless more than 800 times larger than the Sun. The promotion of red giants to red supergiants is largely a nod to how much bigger they are than anything the Sun will ever become.

The size of a supergiant can be a tricky thing to measure, not least because these tremendously large siblings to the Sun also physically pulse, growing and contracting over time. The 1,700 number mentioned above is a rough average size; UY Scuti will exceed or shrink below this value every few years as it goes through its cycle of expansion and contraction. But we shouldn't understate the size of UY Scuti; the distance from its center to its edge is roughly 750 million miles (around 1.2 billion kilometers), which is equivalent to nearly eight times the distance between Earth and the Sun. If UY Scuti were at the center of our solar system, it would extend past Jupiter.

The "edge" of the star is a much harder thing to define for a red giant or supergiant than it is for a star like our own. The surface of any star is defined as the point at which light can stream freely outward into space. But for supergiants and red giants more generally, the star is gradually losing its hold on its outer layers, as they are pressed away and then become so distant that the gravity of the star's core has very little influence over them. Red giants can lose a considerable amount of mass this way, and it means that many of these stars are surrounded by veils of their former surface, forming into their own personal nebula (see the box, opposite). For a star the size of UY Scuti, this nebula of lost gas extends much farther out. In our solar system its cloud of gas would extend out 400 times farther than the distance between Earth and the Sun—10 times farther out than even Pluto's distant orbit.

Stars the size of the Sun won't create a gas cloud quite as large as UY Scuti's—Sol quite simply doesn't have the amount of gas required to spread itself so widely. Red giants can only maintain themselves while they have material to burn, whether that be helium, for stars like the Sun, or heavier elements for more massive stars. When they run out of material, they must find a new resting place. For stars that are less than eight times the mass of our Sun, the result of the next gravitational collapse is the formation of a white dwarf, in what used to be the core of the star (see page 127). The rest of the star, both the diffuse atmosphere, which has already been loosed from it, and the remainder of the star's shells of gas, will be flung outward into glowing, symmetrical clouds of gas we call planetary nebulae.

▲ The red supergiant star Betelgeuse, found in the constellation Orion, is seen here in a color composite image made from exposures from the Digitized Sky Survey 2 (DSS2). The spikes and circular ring are artifacts of the telescope.

▼ **The outer layers of Betelgeuse**

The schematic below shows the approximate scale of Betelgeuse, in orange, relative to our solar system. A vast plume of gas extends outward to a distance so large it would be beyond the orbit of Neptune.

## NEBULAE

A nebula is a cloud of gas, illuminated by the presence of nearby stars. Smaller stars, our own sun included, all meet their ends by producing their own nebula out of gas that used to be their outer layers.

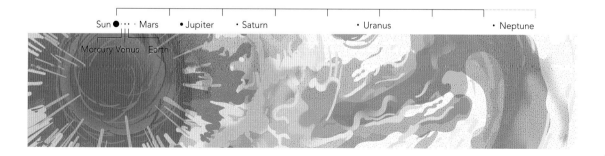

Sun ●··· · Mars   • Jupiter   · Saturn   · Uranus   · Neptune

Mercury Venus   Earth

STELLAR DEATHS

# HOW IS A SUPERNOVA PRODUCED?

We've seen how the members of our stellar family that are on the small side—somewhere less than eight times the mass of our star—will find their end by the last-minute puffing out of a planetary nebula, leaving the hot core of a white dwarf behind. But what of the most massive stars, such as UY Scuti and Betelgeuse, which are much more massive than the Sun?

Once the most massive stars reach the end of their fuel and can burn nothing more, they have a different fate in store for them. They produce some of the most energetic events in the universe—supernovae. A supernova can allow a single star to outshine the entire galaxy's worth of its sibling stars, several billion in number. And they are fascinating events, in part because we can watch them unfold on short timescales by our human standards, after the star itself has existed for billions of years in its hydrogen-burning phase. Supernovae grow in brightness over a period of about a week—a phase called the "rising time." The star will glow its brightest for a few days, and then drop back into dimness over another couple of weeks.

## A SUDDEN COLLAPSE

What happens to a star to make it glow so brightly? Let's go back to the center of our red giant, which in massive stars is burning silicon into nickel. The form of nickel produced in this reaction is unstable, and decays into iron. Once the star runs out of silicon to burn in its core, gravity will again take over, crushing the core down once more. The star can't push back—there's nothing else it can do with the nickel and iron that's built up. This is the same process that generates a white dwarf, but there's a lot more material involved here. If the core of the star is larger than 1.4 times the mass of the Sun, this white dwarf-esque configuration is unstable, and gravity can compress this core more violently and into a denser object. The collapse from something about as dense as a white dwarf to something much more dense happens rapidly, within a couple of seconds. The material in a shell around the core suddenly lurches, and with nothing underneath it, it falls rapidly inward to reconnect with the smaller core of the star. The rapid compression of this shell of material heats it up to a tremendous temperature very quickly, and it releases a huge burst of energy.

Meanwhile, the core of the star stops collapsing, and the free fall of the gas shell is abruptly stopped as well. Coming in with such speed, much of the material bounces away again, in a ricochet outward. Having gathered so much energy, this material is now reflected back squarely at the outer parts of the red giant. It's a one-way path—the sheer energy contained in the rebounding gas blows the star apart, leaving the collapsed core on its own.

What we typically see as the supernova remnant in these images is the shock front of the ejected material hitting any gas and dust that surrounded that star. As the surrounding gas and dust will be a little differently arranged for each star, the shock front will have a slightly different shape for each supernova as well. Depending on the density of the material the shock front is running into, the expansion of the front will continue at different speeds. Inside that edge, there can be a reverse shock, where material has raced up behind the shock front, and bounced off of that back again toward the core of the star that remained behind. This second rebound can heat up the material on the inside of the supernova bubble to very high temperatures.

▲ The Hubble Space Telescope imaged the galaxy NGC 4526, while a supernova (known as 1994D) exploded in its outer regions, visible at the lower left.

The remains of the outer layers of the star, once ejected, are lost to that star. Most of it is simply too far away and moving too fast to collapse back onto the remnant of the core of the star. After getting a speed kick by the rebounding material near the center of the star, the gas that gets flung out of the star moves at a rapid clip. But a supernova is more than just the rapid expulsion of the majority of the star.

## NEUTRON STARS AND NEUTRINOS

During the initial collapse, as the core of the star is pushed past white dwarf densities, the electrons and protons that make up an atom are compressed to such an extent that they form neutrons. Neutrons can sit much closer to each other than protons and their respective electrons, and so there's a new, stable end-point created here, as long as the exploding star is less than about 40 times as massive as our own Sun. This is the genesis of a neutron star. The name is simply descriptive—neutron stars are made entirely of neutrons. They are catastrophically dense cosmic entities. The process of shoving an electron and a proton together also produces an extremely tiny, extremely speedy particle called a neutrino (see the box on page 150).

## ELECTRONS UNLEASHED

On top of the neutrinos flying in all directions, there's another tiny particle zipping around in the aftermath of the explosion—the electron. The material that isn't crushed downward into the neutron star is still made of standard old atoms, consisting of their respective protons, neutrons, and electrons. Typically, the electrons are either bound to a specific atom, or at least floating around in the same material. But with the amount of energy that was flung through the rest of the star, blowing it apart, the electrons can absorb enough of that energy to go their own way. Traveling at close to the speed of light, they can produce X-rays if they are tangled around a magnetic field. The glow of these electrons, moving at fractions of the speed of light, can be detected by X-ray telescopes around Earth.

Generally speaking, it's hard to predict exactly when a given star will cross its internal threshold to begin its collapse and supernova outburst. This unpredictability is one reason we like the neutrino alarm mechanism—if we detect a burst of neutrinos, we have a little bit of a heads-up that something is coming. Our best method of

spotting lots of supernovae is still to survey the sky repeatedly and look for changes—something that upcoming telescope facilities like the Vera C. Rubin Observatory will excel at. Occasionally, however, we find a nearby sibling of the Sun that looks like it might go off soon (on an astronomical timescale, of course). Such is the case with Betelgeuse.

▶ **Top:** The cloud of gas left behind by supernova E0102 is seen in this Hubble image as filaments of blues in the lower half of the image. At the top right is an expanse of new star formation, with the characteristic pink glow of hydrogen—the entire life cycle of stars in one image.

**Bottom:** An illustration of the structure discovered around Supernova 1987A, where two rings of material were ejected from the exploding star. This structure was surprising, as it indicates the star did not explode symmetrically. Why that might be is still a mystery.

## NEUTRINOS

A neutrino is a fundamental particle of our universe. It is neutral, as the name implies, and has extremely little mass. Neutrinos very rarely interact with matter (other than via gravity); they travel the universe at very close to the speed of light, and they're formed in a number of places. Our own star forms some neutrinos in its core. Because the neutrino doesn't interact very much with the stuff of stars and planets, once formed, a neutrino is very likely to escape the star almost instantaneously—unlike the poor photon, which will spend hundreds of thousands of years wandering the radiation zone.

Neutrinos may be generated by our Sun, but they're produced in much larger quantities during the collapse of a supernova and creation of a neutron star. Their instantaneous departure from the star means we on Earth can use them as an early warning system for nearby supernovae. The light may be still trapped within the star during this early collapse, as the rest of the star hasn't been blown outward yet, but the neutrinos have already escaped. The time delay before the light can escape is not necessarily very long—for Supernova 1987A, it was only a few hours.

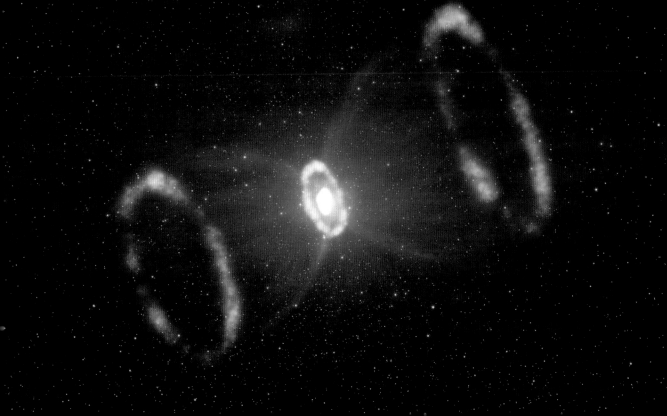

# HOW BRIGHT IS A SUPERNOVA?

Betelgeuse is a red supergiant star, and we've mentioned already that it's the left-hand shoulder of the Orion constellation, a visibly orange star in the night assembly. It is, in fact, one of the top 10 brightest stars in the night sky. (The Sun is somewhat obviously always the brightest object in the daytime sky.) Betelgeuse is also one of the few stars that's close enough for us to resolve in more detail than a point source of light.

At a distance of 600 light-years from Earth, if Betelgeuse were to go supernova right now, we'd clearly have a 600-year delay before we could spot any changes from here. Whenever it explodes, it's going to be a long wait before we see it, but in the meantime we can do some rough maths to work out how bright Betelgeuse's explosion will be. The precise flavor of this supernova is still up for a bit of debate, depending on exactly how fast it's spinning and how quickly it gets rid of its outer layers over the next 100,000 years. But all the supernovae options for this star reach about the same brightness, so for a quick calculation we don't really need to worry about the exact type of explosion this star will undergo.

## GAUGING BRIGHTNESS

There are two ways of measuring brightness in the astronomy world. The first is absolute magnitude, which is the brightness of the star as it would be measured from a fixed distance. (It's arbitrary, but the fixed distance chosen is 10 parsecs, or about 33 light-years.) This is trying to get to a measure of intrinsic brightness—as though we could line up all the objects in the sky at equal distance from us, and compare them to each other. We can't actually measure the brightness of a star this way, but we can apply some corrections based on the distance to the star in order to get to it. The absolute magnitude of a Type II supernova is around –17. Because astronomers have the worst conventions in the world (for largely "historical reasons"), negative numbers mean brighter objects. The Sun has an absolute magnitude of 4.83, which, once we

translate out of "magnitudes," means that the Sun is 500 million times fainter than the –17 supernova, when measured at the same distance. This huge difference in relative brightness shows why a supernova can outshine an entire galaxy.

The other method of measuring brightness is a bit more straightforward. It's the apparent brightness—how bright it appears to us as viewed from Earth. In this frame of reference, more distant objects will always appear fainter, regardless of how intrinsically bright they are. Because Betelgeuse is still fairly distant from us, the apparent brightness will be significantly less than the absolute magnitude. Based on the distance to Betelgeuse, we can work out that the apparent magnitude of the peak of the explosion will be –10. The Sun, the brightest thing in our sky, checks in at an apparent magnitude of –26.74. Once again, translated out of magnitudes, this means that the Sun, as seen from Earth right now, is a whopping 5 million times brighter (more or less) than Betelgeuse's explosion, so this supernova certainly won't be anywhere near as bright as our Sun in the daytime. That's not to say you wouldn't be able to see it—it would definitely be bright enough to see during the day, as long as you were looking in the right direction. (After all, as we noted before, you can still see Venus in the daytime if you know where to look!)

Nighttime will be a different story. The brightness of Betelgeuse's supernova will be about the same as the quarter Moon. It will also be about 16 times brighter than the brightest supernova known to

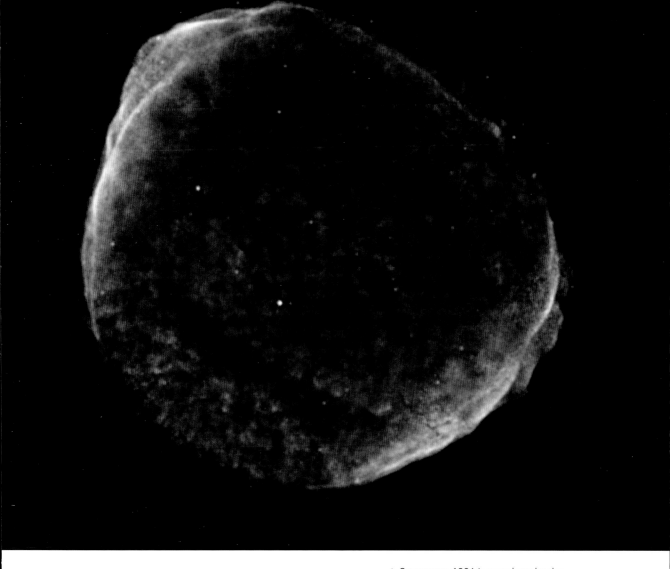

have been seen from Earth, which occurred in the year 1006 and was recorded by a number of early civilizations. It was said that the supernova in 1006 was bright enough to cast a shadow at night. Betelgeuse, being significantly brighter, will likely also cast shadows—which, if you think about the brightness of a quarter Moon, would make sense.

Betelgeuse isn't expected to explode for another 100,000 years or so, but we do expect a few supernovae in our galaxy every couple of hundred years, so there are a number of stars that are very near the ends of their lifetimes within the Milky Way.

▲ Supernova 1006 is seen here by the Chandra X-ray observatory; red in this image corresponds to the glow of gas heated to millions of degrees, and blue to the ultra-fast motion of tiny particles.

STELLAR DEATHS

# WHAT DOES IT MEAN WHEN STARS ARE "TWINNED"?

We mentioned earlier that not all stars are as isolated as the Sun. Stars can gather together in tight globular cluster families, but there's another common configuration—binary stars. These are stars that formed close enough to each other that they are gravitationally tied together, and will travel the length of the galaxy like this. These stars are twins, of a sort, though most of them will be fraternal, not identical twins.

About a third of the stars in the Milky Way are in some kind of binary, though this accounting includes all the stars that are only very gently tied to each other. If you want to include only close binaries, where the two stars orbit each other in years instead of hundreds of years, the number drops down considerably. Since the majority of these stars aren't identical twins, they won't have exactly the same mass. If their masses are different, one star (the more massive of the two) will turn into a red giant before the other. It will expand outward into the space between the stars, and the second star, which is still burning hydrogen in its core, may begin to tug on the outer layers of its red giant twin, slowly pulling them in toward itself. The red giant, feeling this pull, will no longer look like a giant fuzzy sphere, but the side of the star facing its twin will be pulled away sharply into more of a fuzzy teardrop shape.

What the companion star does with that surface material depends on what kind of star it is. If the companion to the red giant is still burning hydrogen in its core, it should be relatively stable on its own. But with an influx of gas from its red giant twin, the stable twin will find itself unexpectedly growing in mass. If you grow a star that's already able to burn hydrogen, you've added to its gravitational inward force, meaning that the temperature at the core of the star can be increased, and the speed at which it can burn through its hydrogen also increases.

What of the red giant? Well, it was going to lose those outer layers anyway—donating them to its twin is the fastest recycling job it could have done. And it won't donate itself into oblivion. Its twin star can only pull the layers that are furthest from the red giant's core; the hot dense core of the giant is staying right where it started. This setup can last for a long time; the red giant, slowly expanding, gradually feeds more material directly into the gravitational weight of its twin star. That twin star, by growing in mass, only shortens its own lifetime, burning hotter and faster than it would otherwise, had it not had a twin star to keep it company.

> The red giant, slowly expanding, gradually feeds more material directly into the gravitational weight of its twin star.

▶ The Kepler 16 star system involves a planet (Kepler 16b) orbiting a pair of stars, both of which are smaller than the Sun. Spotted through a cosmic coincidence, the Kepler space telescope is able to see the two stars pass in front of and behind each other, as well as the planet crossing in front of both of them, leading to the unlikely alignment illustrated here.

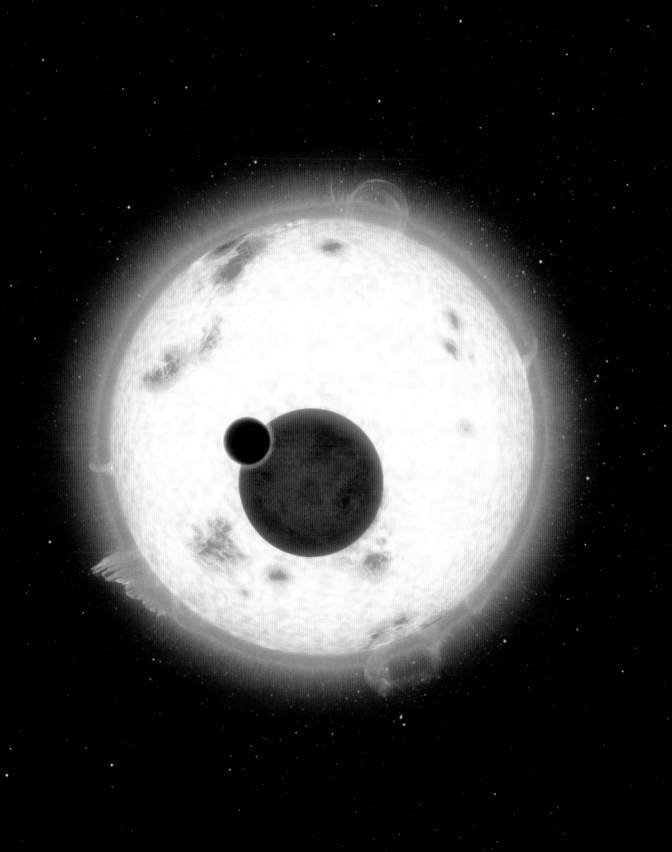

# WHAT HAPPENS AFTER A STAR EXPLODES?

When we dust off our post-red giant bursts (see pages 142–143), we're largely left with low-mass stars that have produced planetary nebulae and remain as white dwarfs, and high-mass stars that have turned into supernovae, leaving only neutron stars behind. Both white dwarfs and neutron stars have been found in pairs with other, evolutionarily younger stars, a role-reversal of our fraternal twins.

In these situations, instead of a red giant and a hydrogen-burning star, we have a hydrogen- or helium-burning star and a white dwarf or neutron star. Let's start with a red giant, the Sun's helium-burning stellar sibling, paired off with a white dwarf. Much like the scenario we outlined earlier, the white dwarf, if it is close enough, will pull material away from the red giant, gradually adding to its own mass. Unlike the hydrogen-burning star, though, a white dwarf is doing no burning in its core. All that happens is that its mass grows.

White dwarf stars are not terribly tolerant of gaining weight. Eventually, this white dwarf will trigger an explosion of its own. The less destructive option is for a thermonuclear detonation to occur on the surface of the white dwarf. This surface detonation is called a "nova." Like the supernova, it is a dramatic, bright flare; unlike the supernova, it is not so bright that it will outshine the galaxy it lives within, and it doesn't destroy the white dwarf. So the white dwarf, with a steady source of fuel,

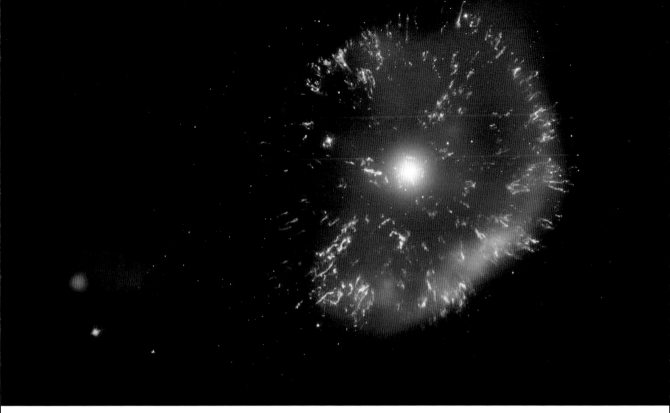

▲ This image of GK Persei shows the aftermath of a colossal white dwarf nova explosion. This image is a composite of X-ray light in blue, visible light in orange, and radio in purple. At the time of its explosion, it was briefly one of the brightest stars in the night sky.

◄ An illustration of the RS Ophiuchi binary system, which contains a white dwarf star (right) and a red giant (left).

can build up enough mass to trigger a detonation, blasting the material away from its surface, before settling down to build up more mass, triggering yet another surface explosion. This kind of behavior makes these stellar twins fairly noticeable, because the brightness of the star will repeatedly flare to many times its original level.

The other option is for the white dwarf to undergo a supernova. The white dwarf itself is only stable up to a mass of 1.4 times the mass of the Sun. Beyond that, the star is gravitationally compressed beyond what it can stably support. The gravitational crushing forces for a white dwarf can very briefly press the star to a temperature where it can burn carbon. And so it does, catastrophically, and all at once. The energy unleashed by the whole star undergoing a giant simultaneous burn flings the star apart, leaving nothing behind. This style of supernova is unique to white dwarfs—the higher-mass stars have already burned through all the carbon at their cores.

STELLAR DEATHS

## X-RAY BINARY

If, instead of a white dwarf, the twinned star is a neutron star, we wind up with what's called an X-ray binary, for the somewhat boring reason that it produces a lot of X-rays. As the gas from the red giant (or a hydrogen-burning star) is pulled away, it gets stretched into a very thin disk surrounding the neutron star. The disk forms because it's very hard for gas to lose a lot of momentum all at once and plunge straight down onto the neutron star. Because the disk is there, the gas is heated up to an incredibly high temperature before it makes it all the way to the neutron star—or to a black hole (see pages 160–161). This heat causes the X-ray glow we can observe, and keeps the disk itself almost invisible in optical light.

No matter what kind of supernova or nova explosion a star undergoes, these catastrophic, extremely energetic events have no impact on the galaxy as a whole. Even if a star entirely self-destructs, it only makes up one trillionth of the mass of its galaxy. That explosion, even though it can outshine the galaxy, is simply not big enough to have a noticeable impact—one star can only put out so much energy. The outshining can be dramatic in images of the galaxy, however—a bright star can suddenly appear, and then gradually fade from view.

There is a critical role for supernovae in a galaxy, though. Without the supernova, we'd only have elements up to iron to work with. During the ricochet of the outer core of the star, the heavier elements are formed, giving us gold, silver, and platinum, for instance. These elements are scattered along with the remains of the star into space. When the next generation of stars forms, they will incorporate those metals into themselves, carrying forward the remains of generations past. Earth's supply of these precious metals, along with all the other elements on the periodic table beyond iron, come from previous generations of our Sun's siblings catastrophically exploding and spreading these elements throughout the galaxy.

> Earth's supply of these precious metals, along with all the other elements on the periodic table beyond iron, come from previous generations of the Sun's siblings catastrophically exploding and spreading these elements throughout the galaxy.

There's one outcome of a supernova we haven't looked at yet. If the star is massive enough, the collapse of its core won't stop at a neutron star. Gravity will continue to pull each neutron closer to its neighbors. Unfortunately, there's no new particle that neutrons can collapse into when compressed too far, and what happens next is mathematically difficult to describe. At this point, the runaway collapse should continue. With no resistant force it can become infinitely small, with an infinitely deep gravitational well. This object, infinitely small, infinitely dense, is termed a singularity. Infinities and physics currently do not play well together, so our descriptions of how space and time behave immediately surrounding the singularity begin to fail. (The singularity itself is currently most accurately described by a large number of question marks.) This collapse has produced a new astrophysical object—it has become a black hole.

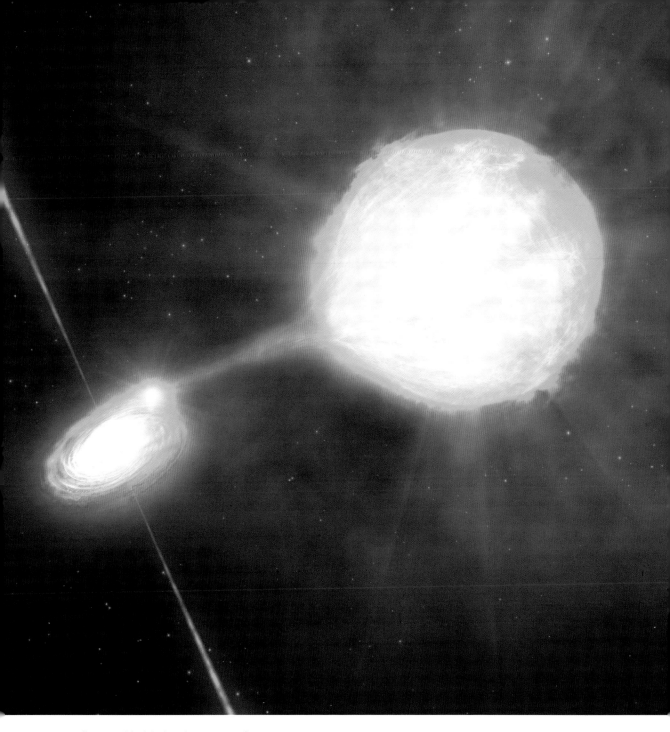

▲ A stellar mass black hole orbiting around a massive blue giant, which will eventually become a black hole itself. The black hole this image is based on, which is located in the spiral galaxy NGC 300, is slowly peeling away the outer atmosphere of its companion, and building its own accretion disk.

# HOW DO WE KNOW BLACK HOLES REALLY EXIST?

Black holes are some of the most interesting remains of the Sun's sibling stars, and they were first suggested by people who were tinkering with our understanding of how gravity works. Black holes, at the time, were completely theoretical additions to our cosmic family. In 1784, John Michell realized that if you had a sufficiently compact, massive object, the escape velocity—the speed needed to escape from the surface and make it out into orbit and beyond—would be faster than the speed of light. Since nothing moves faster than the speed of light, these objects become a trap for anything that comes close enough to get dragged in.

The term "black hole" has expanded a bit to mean more than just the very "object" of the black hole, that infinitely small singularity. This is reasonable enough, because there is a region of space surrounding the black hole where the gravitational pull of the singularity is so strong that light itself would be captured, which is marked by an imaginary line called the event horizon. This event horizon is usually quite distant from the actual singularity, but the term "black hole" is often slung around for this entire region, because it is inside this space that the black hole's reign is complete.

## FINDING PROOF

It takes a lot of evidence to get us from a theoretical member to a confirmed member of our stellar family, but fortunately over the years we've built up just such a pile of evidence for their existence. Some of the best evidence we have for adding black holes to our cosmic family tree comes from the center of our own galaxy. With the telescopes we have now, we can watch individual, luminous stars zip around some very heavy, physically small and completely invisible object. From the orbits of the stars, we can figure out how much matter the dark object must contain. We also know how big it could be, since some of these stars pass very close to the mystery object,

and aren't being torn to shreds, so it must be smaller than that. With the mass and the size, we can have a guess at its density.

The object at the center of the Milky Way is so dense that no other known or suggested object from anywhere in our family tree, aside from a black hole, can explain our observations. The object is about 4 million times the mass of the Sun in a space that's less than the size of our solar system. The only way you can pack material that tightly is if you crush it to the point where the object has no other physical option than to be so dense that light cannot escape it: it must be a black hole.

We know that other galaxies have black holes too, so ours isn't weird or special in some way by having this massive black hole at its center. In fact, as far as we've been able to tell, every galaxy has a black hole at its center. If you're looking at a galaxy that's relatively close, you can tell how big its black hole is by how the galaxy is rotating. If you want to be able to model the way it rotates, you need to have a good understanding of where the matter in the galaxy is. Generally we can find out where most of the matter is, because a good chunk of it is either in gas (which we can see) or stars (which are even easier to see). However, if we don't also include a whole pile of extra matter right at the center (like

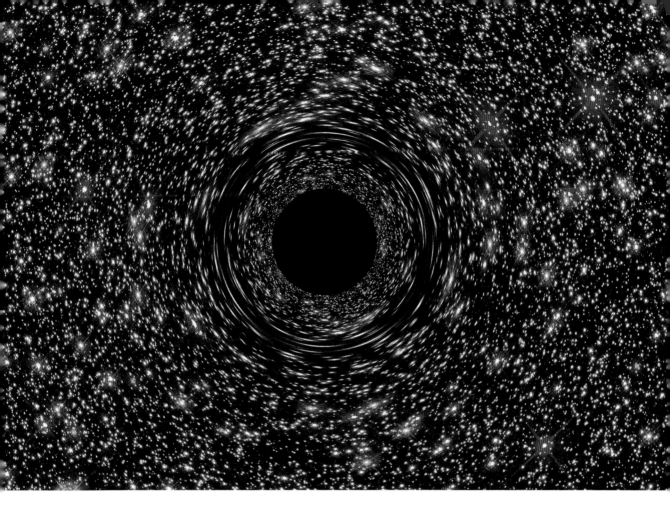

a black hole), the models won't reproduce what we observe. If we stick a black hole in there, we get a much better match.

Similarly, we know that there are smaller black holes hanging around in our galaxy—supermassive black holes aren't the only ones in the universe. Much like we see stars zipping around an invisible, apparently massive object in the core of our galaxy, we've seen a huge number of stars that appear to be orbiting an invisible companion closer to their own mass—the twin stars arrangement again, but instead of a white dwarf or a neutron star, their companion is entirely invisible. We can repeat the density calculations we did before, and we find that, again, we're dealing with an object that is sufficiently tiny and sufficiently massive that only a black hole fits the requirements. Our family has a new member—the black hole.

▲ This computer-simulated image shows what it might look like to see a supermassive black hole at the core of a galaxy. Light from the stars behind the black hole would be stretched out into long arcs; but the black region in the center represents the black hole's event horizon, where no light could reach us.

STELLAR DEATHS

# WHY WON'T BLACK HOLES EAT THE UNIVERSE?

Black holes are a particularly imagination-inspiring member of our cosmic family, but their role as a light-trap, along with the difficulty of observing them directly, has created some myths about them. One of these is the idea that a black hole is some kind of universal-scale vacuum cleaner, constantly sucking material inward toward itself.

This point of confusion seems to come from a few sources. One is that we often describe the force of gravity as one that "pulls" or "attracts" two objects together. If you combine that image with the idea that a black hole can be considered like a cosmic trash can (throw anything in, nothing comes out), then it seems very reasonable to think that not only are they inescapable light traps, but that they pull things toward them as well. Unfortunately, black holes do no pulling whatsoever, as we'll see.

### 3D DISTORTIONS

When we talk about the gravitational influence of an object, it's often described in terms of bowling balls on rubber sheets. The more massive the object, the deeper the indentation in space becomes, and the steeper the indentation finds itself.

These are good and reasonable explanations, though they are simplified since the indentations happen in all directions, so these "indentations" are really three-dimensional distortions, or condensations, of space itself. These simplifications do illustrate an important point—fundamentally, gravity operates like a distortion to space. Often we speak of this distortion in terms of a "well," which goes back to this idea of a two-dimensional sheet bent into a third dimension. It's easy to think of things rolling downhill into a divot, formed by the presence of a large amount of mass.

Mathematically, the influence of gravity is written out as directly proportionate to the mass of the object, and inversely proportional to the square of the distance between you and that object. The direct dependence means that as you increase the mass, you increase the influence of gravity by the same amount. The inverse dependence means that as you double the distance, the influence of gravity gets four times as weak. This strong dependence on distance is what creates the particular smooth curve that we draw to illustrate gravity's influence from the center of the object, outer parts of the diagram to the right. If you're very close to the massive object, the gravitational well is deep, and you feel a strong gravitational force. If you're farther away, gravity can't distort space very much, and so you feel a much weaker gravitational force.

Mathematically, the influence of gravity is written out as directly proportionate to the mass of the object, and inversely proportional to the square of the distance between you and that object.

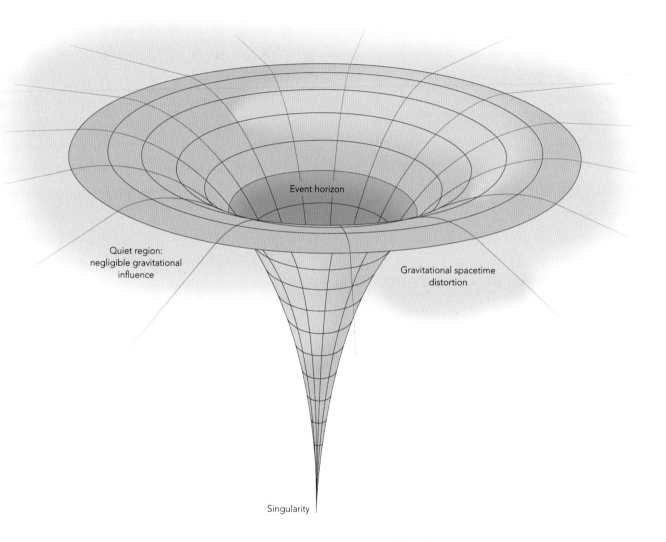

Quiet region:
negligible gravitational
influence

Event horizon

Gravitational spacetime
distortion

Singularity

### ▲ Black hole regions

The "rubber sheet" model for a black
hole. At large distances from the black
hole, the influence of the black hole is
very weak. It's only when you get close to
the event horizon of the black hole that
the distortion of space starts to become
difficult to escape from. Deep within this
distortion is the black hole singularity.

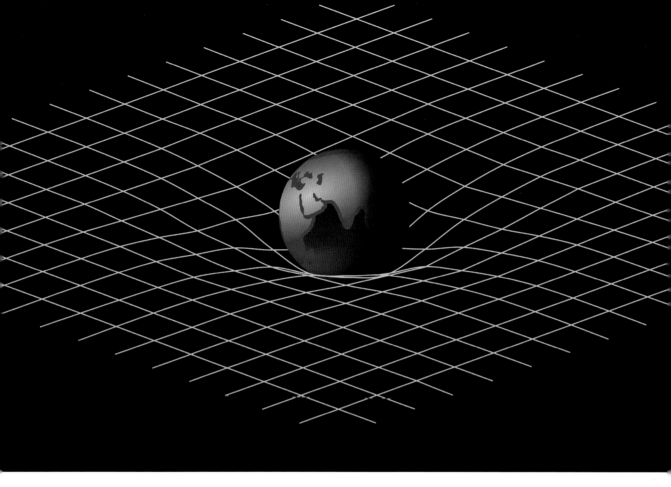

## GRAVITATIONAL WELLS

All objects in the universe create their own distortions to space, including our human bodies. However, the force of gravity is relatively weak, and we humans are not very massive, so we don't expect to see objects around us rolling into our gravitational well. Once you scale up to moon or planet amounts of mass, then the force of gravity means they can have a little more influence on their surroundings. Each of the planets thus has its own distortion to space, but all of these distortions pale in comparison to that caused by the presence of the Sun.

Critically, all of the distortions caused by massive objects didn't spontaneously appear when the object finished forming. Each object had a gentler gravitational well before it finished collapsing into a denser object. A small gas cloud could have the same mass as a small planet, but the planet's

▲ For an object that is much less dense than a black hole, such as Earth, the distortions caused to the fabric of space are milder, though still present and measurable.

## GRAVITATIONAL POTENTIAL

Gravitational potential is a way of quantifying how much energy you would have to expend in order to lift something away from a massive object. The deeper you are in the well, the more energy is needed, and the steeper the well, the faster you'd need to go to escape on a rocket.

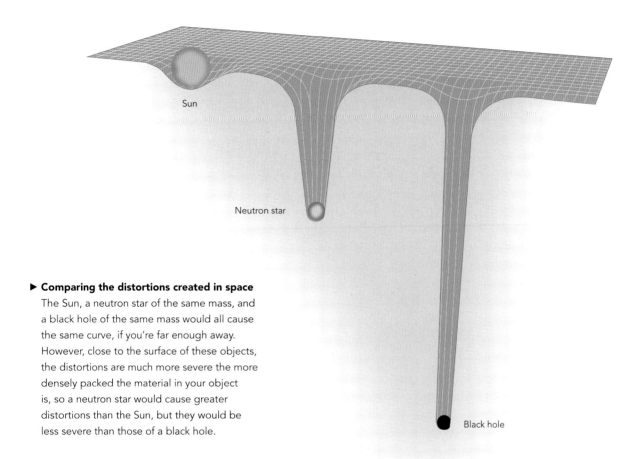

Sun

Neutron star

Black hole

▶ **Comparing the distortions created in space**
The Sun, a neutron star of the same mass, and
a black hole of the same mass would all cause
the same curve, if you're far enough away.
However, close to the surface of these objects,
the distortions are much more severe the more
densely packed the material in your object
is, so a neutron star would cause greater
distortions than the Sun, but they would be
less severe than those of a black hole.

gravitational well will be steeper in the center
than the gas cloud's. If you kept compressing
the planet's mass until it was dense enough to
be considered a black hole, its gravitational well
would be steeper yet in the center. The steeper
the gravitational well, the faster you need to go
to escape it.

The event horizon of a black hole, in this context,
describes the boundary within which the black
hole's gravitational well is so steep that not even
light could escape it. If you take one of the rubber
sheet diagrams, you could place down a circle to
describe the event horizon's location, though in
the universe these are actually spheres in three
dimensions. The event horizon itself doesn't
describe a physical boundary to the influences
of gravity. The gravitational distortion exists
continuously outside and inside of the event
horizon—outside the event horizon it's just slightly

less steep. You could place another circle outside
the event horizon that describes the location at
which you need to go half the speed of light to
escape—again, there's no change or boundary in the
physical distortion at this circle, it's just a descriptive
line that might prove useful to understand or
describe the object's influence on space.

If you're far enough away from the black hole, then
the influence of gravity from that black hole is no
different from having a star of the same mass—or
a planet, if you somehow got yourself a particularly
tiny black hole. If you replaced our Sun with a black
hole of the same mass, the planets would continue
to orbit the way they always have; their orbits arc
determined entirely by the mass of the object at
the center of the solar system, and not by that
object's density. If our grandparent star couldn't
pull an object into its atmosphere, the black hole
would struggle to do so as well.

# WHAT SHAPE IS A BLACK HOLE?

If we use the event horizon of the black hole to describe the "surface" of the black hole, even though there's no surface there, we can start to describe the geometry of this sibling of the Sun. We mentioned earlier that the event horizon is usually a circle when we simplify the gravitational pull down to two dimensions, so in three dimensions, the event horizon should be a sphere. In fact, black holes are probably some of the most perfectly round objects in the universe.

In principle there's no up, down, or sideways to a black hole, just as there's no up, down, or sideways to a spherical toy. If the black hole is totally isolated, this is the case. Any direction you look at it, it will still appear as a circle of utter blackness, where no light is produced, reflected, or passing through. The only change you'd see by looking at it from different angles would be the backlighting. The backlighting of the black hole isn't just the background light; black holes bend light that passes near them, and so all the light from the stars behind the black hole can end up compressed into a ring of light, surrounding the black hole itself. The more objects behind the black hole, the brighter that ring of light will appear. If you found a direction with very few stars behind the black hole, this ring of light might seem to vanish.

## A MANY LAYERED BLANKET

The exception to this is if our black hole is one of a set of twins, alongside a companion star. Much like the white dwarf star and the hydrogen-burning star, the black hole will start to siphon gas off of the surface of the other star, if the star's atmosphere expands outward and drifts too near. While it's one thing to pull the gas away from the other star, it's quite another to pull that material inward past the event horizon of the black hole. The gas from the other star has to lose a lot of energy in order to fall into the black hole, and so the black hole winds up wrapped in a many layered blanket of the other star's gas. This is called an accretion disk, as the many windings of the other star's gas form into a disk shape around the black hole. Accretion is simply the black hole's attempt to gain that material for itself, and grow its own mass in turn. Black holes are very bad at getting material that's handed to them like this all the way down past the event horizon, so the accretion disk is usually relatively stable, and the black hole doesn't actually grow in mass very much.

> Any direction you look at it, it will still appear as a circle of utter blackness, where no light is produced, reflected or passing through.

However, thanks to the presence of this accretion disk, there's suddenly a direction, or an orientation to the black hole. You could look at the black hole "from the side," which would be looking at the edge of the accretion disk. Or you could look down on the disk "from the top," where you'd see the black hole, plus the glowing disk of gas surrounding it.

There are a few other situations in which you might see some weird things around a black hole, and where the spherical symmetry is broken. One of them is if two black holes are orbiting each other. You could get this scenario from a twin star setup, where both of the stars were massive enough to

▲ The accretion disk surrounding this artist's illustration of a black hole in a binary system with another star is incredibly bright due to its high temperature causing the gas to glow. Here we see the accretion at a slight angle from above, which obscures the black hole in the glare from its accretion disk.

explode in a supernova and create a black hole in their aftermath, but without throwing their twin out of its orbit. Light bends around a single black hole in a pretty dramatic fashion; if you add a second black hole in close proximity to the first, the dance that light can get caught in is an extremely elaborate one. This convoluted dance means that only one of the black holes actually looks spherical; the other one's shape is distorted, as the ring of backlighting is also distorted, and the black hole nearest to us ends up with a small curved arc of darkness hovering near it. This has technically been called an "eyebrow." They're also referred to as "black hole shadows," which is just as great, name-wise.

# WHAT HAPPENS WHEN BLACK HOLES COLLIDE?

Twinned black holes are quite stable, and they are capable of lasting billions of years without much change to the way they orbit each other. However, if their orbits are not very circular or the black holes are not the same mass, or if enough time has passed, you can catch them colliding. We had hoped for many years that we might find a few whose orbits had fallen toward each other. Fortunately, new technologies now enable us to catch the signature of collisions between black holes.

**B**lack holes go through a long spiraling wind-down before they actually collide, and the distortions they make in space as they circle each other create gravitational waves in space-time, similar to swirling your finger in a pond. The formation of these waves carries energy away from the two black holes, helping them lose enough energy to merge together. Like sound waves, gravitational waves have a frequency and amplitude. These are at extremely long wavelengths, so it's nothing the human ear could hear normally, but if you scale the frequencies up by a couple of thousand, you can get an impression of what it would sound like. This scaled sound is often described as a chirp, but if you listen to one, it sounds much more like a *vwooop*.

Making these noises isn't just a curiosity—there's a real scientific reason to create them. As the Laser Interferometer Gravitational-Wave Observatory (or LIGO for short) goes hunting for the signature of gravitational waves in the universe, the collisions between black holes are one of the first things we have been able to detect—they're one of the "louder" sources of gravitational waves. As with any detections, it helps to know what you're looking for, and these predictions of what black holes of a specific size should do to the space surrounding them as they merge are key to developing automated methods of spotting these gravitational waves as they wash over our planet.

On February 11, 2016, LIGO announced the very first detection of the gravitational wave signature of two black holes colliding. With a large set of predicted *vwooops*, the LIGO team found that their particular chirp was closest to a model of two black holes, one 36 times more massive than our Sun, and one 29 times more massive, reaching the end of a long spiraling orbit, and finally colliding.

After the initial strike of its twin black hole, the conglomerate black hole continues to vibrate for a long time, gradually losing energy to space until it finds itself back the way it started.

When they merge, the event horizons of the two black holes cross and we can consider them as a single object. After all, there's now one contiguous region where light can no longer escape. This is another moment where black holes are not perfect spheres. The newly formed larger black hole will quickly become reasonably round-ish, but it has to go through a fairly long process called "ringdown" before it can completely settle back down into its earlier, perfect sphere shape.

If ringdown sounds like a melodic term to you, you're spot on—it's meant to evoke the gradual dimming of sound that comes from a bell after being struck. The bell continues to vibrate long after the initial strike, carrying a tone at a quiet level much longer than you might think. So it is with our Sun's black hole cousins. After the initial strike of its twin black hole, the conglomerate black hole continues to vibrate for a long time, gradually losing energy to space until it finds itself back the way it started—an isolated sphere marking a region of gravity's absolute dominance.

▲ This image of NGC 6240 shows an optical image from the Hubble Space Telescope. This galaxy, which has recently collided with another galaxy, contains two supermassive black holes, expected to eventually merge into a single, even more supermassive black hole.

# WHAT IS TIME DILATION?

Black holes are also rolled out whenever scientists wish to discuss something odd that happens when space or time behave in an unusual fashion. It's the bending of light in strong distortions of space that gives black holes their odd backlighting, but as space and time are bound together as space-time, whenever space is doing something odd, time must be too.

Within Einstein's general relativity, there is a concept called time dilation, which implies that our perception of how quickly time is passing depends on a combination of how fast we're moving and how distorted the space in which we are sitting is. Space is distorted near very massive objects (like black holes) and so if we're either moving quickly (relative to the speed of light) or near a black hole, our perception of time might differ from the perception of time that someone else, who's not moving quickly or near a black hole, would measure.

Movies like *Interstellar* brought the idea of time dilation a little farther into the public awareness, and in actual fact they did a pretty good job of explaining what was happening, especially considering how complicated anything to do with relativity can get. Time dilation is a property of our universe, though, and it's apparent on much smaller scales than the black hole of the movies—we can easily measure its role around Earth. If you've ever used the "my location" option on your favorite map app on your phone, you've made use of it yourself.

## SLOWING DOWN

The principle of time dilation is this: the deeper you find yourself in a strong gravitational field, the slower your clock will run, relative to someone who is not in as strong a gravitational field. This can be translated into meaning the closer you are to something large, the slower your clocks will run. However, it also means that if your two clocks are the same distance from two objects, one which has a much stronger gravitational pull than the other (say, a planet for one clock, and a black hole for the other), the clock around the more massive object (the black hole) will run slower, even though both clocks are the same distance away from their respective objects.

A GPS satellite, which orbits about 12,500 miles (roughly 20,100 kilometers) above the surface of our planet, is clearly farther away from Earth than those of us who live on its surface. A clock on Earth's surface will therefore appear to progress more slowly, relative to the clock on the GPS. Unfortunately, we really need these two clocks to operate in sync with each other, otherwise the location information you get back from the satellite starts to be increasingly inaccurate. This effect can be mathematically calculated, so we can correct for this slight difference in clock speed by setting the GPS clock to run a little slower than normal. Just a little slower—near Earth, the difference is a matter of nanoseconds.

◀ The *Solar Orbiter* spacecraft is illustrated here on a close flyby of the Sun. Traveling so close to a massive object like the Sun, this craft's clock will have to take into account issues of time dilation.

▼ **GPS technology**
The GPS system receives information from GPS satellites overhead, each of which broadcast the time and their location. With both pieces of information, the GPS on the ground can figure out the distances to each of those satellites, and therefore where it must be, between them all.

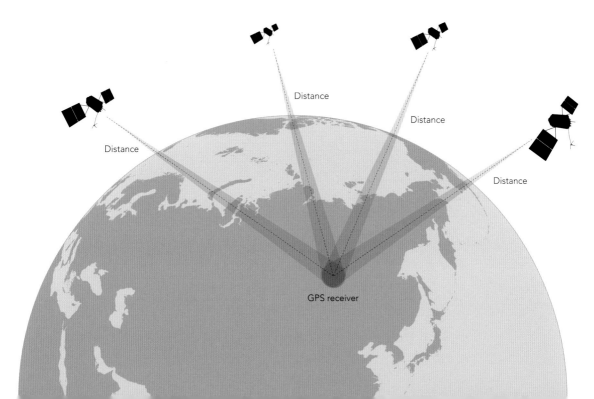

Distance

Distance

Distance

Distance

Distance

GPS receiver

But for a GPS, another form of time dilation is also in effect, because the fact that the satellite is in motion changes the onboard clock as well. This form of time dilation is such that the faster you're moving, the slower your clock appears to move, relative to someone who is not moving. This effect gets more and more dramatic the closer to the speed of light you travel, so the offset between your clocks would get more severe.

A GPS satellite, which has to travel in a circle 12,427 miles (20,000 kilometers) above Earth's surface every 12 hours, is therefore going around 8,670 mph (14,000 kph). The speed of light is about 670 million mph (around 1.07 billion kilometers), so our GPS is only going about 13 millionths of the speed of light. This is not a significant fraction of light speed, but it's enough that we can calculate and measure the impact that it has on the onboard clock on the GPS.

As it happens, the gravitational time dilation has a more significant impact on the clock than the speed of the GPS relative to us on the ground. While the fast running of the clock due to gravity is partly canceled out by the slowdown imposed by its 8,670 mph speed, the clock is still running a bit fast because it's so far away from the most concentrated part of Earth's gravitational field. It's this slightly diluted rapid ticking of the clock that's calibrated into the clocks that we send up onto our satellites for the GPS network.

If you're in a lower orbit, as the astronauts on board the International Space Station (ISS) are, the gravitational effect isn't a big one, so the only clock-altering effect they deal with is due to their orbital speed around our planet. If you spent an entire year on the ISS, like astronaut Scott Kelly, you would have aged a grand total of 0.007 seconds less than your family on the ground. This is an extremely small number, not because the ISS is going particularly slowly (it orbits at around 4.8 miles/s, or 7.7 km/s), but because the important factor is how close to the speed of light you're going, and for the ISS, the speed of light is so much faster—about 39,000 times faster, in fact.

Anytime you pull your phone out to get directions to a restaurant, or to check how far you have left to walk, and your phone accurately figures out where you are, the GPS connection that underlies that software is relying on our understanding of relativity, and the time dilation the satellite is undergoing, to do its job properly. The deeper and steeper the gravitational well, the more important these accountings of time dilation are, so if we were exploring a black hole or a neutron star, these corrections would be critical!

## ORBITING A BLACK HOLE

If you were to put yourself in orbit around a black hole, where the gravitational distortions are much more extreme, all of these effects would also be much more extreme. An orbiting craft around a black hole, if the black hole were large enough, could absolutely feel time running considerably slower than a spacecraft that hadn't approached anywhere near it.

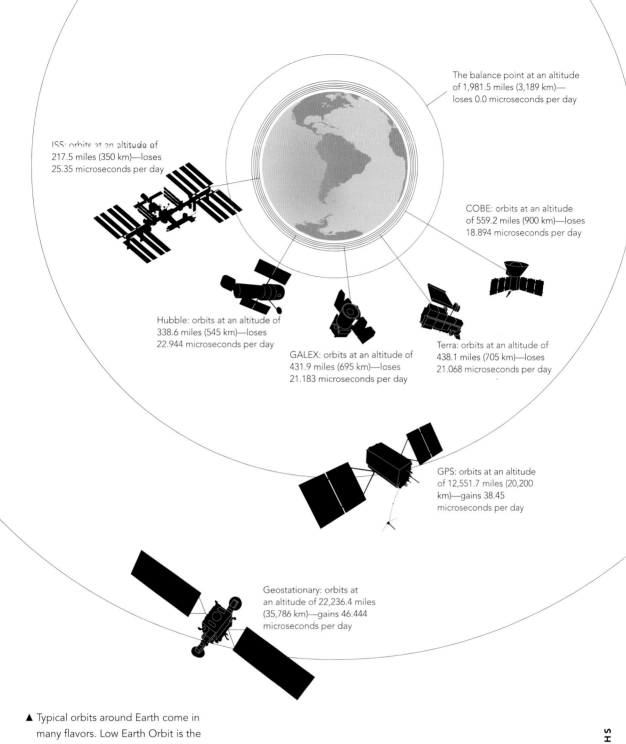

The balance point at an altitude of 1,981.5 miles (3,189 km)—loses 0.0 microseconds per day

ISS: orbits at an altitude of 217.5 miles (350 km)—loses 25.35 microseconds per day

COBE: orbits at an altitude of 559.2 miles (900 km)—loses 18.894 microseconds per day

Hubble: orbits at an altitude of 338.6 miles (545 km)—loses 22.944 microseconds per day

GALEX: orbits at an altitude of 431.9 miles (695 km)—loses 21.183 microseconds per day

Terra: orbits at an altitude of 438.1 miles (705 km)—loses 21.068 microseconds per day

GPS: orbits at an altitude of 12,551.7 miles (20,200 km)—gains 38.45 microseconds per day

Geostationary: orbits at an altitude of 22,236.4 miles (35,786 km)—gains 46.444 microseconds per day

▲ Typical orbits around Earth come in many flavors. Low Earth Orbit is the easiest to reach, and is where the International Space Station resides. GPS and Geosynchronous orbits are much farther from the surface of Earth. The Hubble Space Telescope orbits in a relatively low orbit, but about 110 miles (180 kilometers) farther up than the ISS.

STELLAR DEATHS

# What happens to time dilation at the speed of light?

Time dilation is present for all moving objects, and the faster you move, the stronger it is. How should it affect light itself? After all, light is the fastest thing in our universe. The principle of time dilation tells us that the faster a voyaging craft is going, the slower its clock will appear to tick to an outside observer not moving at speeds that are substantial fractions of the speed of light. The voyagers themselves see a normal clock inside their craft, but clocks outside their craft appear to be running slow. So if you put one person on a very fast spaceship and send them to go explore the cosmos, and keep another person at home to wait for their reports, the faster the spaceship goes, the more discrepant their clocks will be.

Getting to grips with this time dilation problem will be a two-step process. We first need to understand more about exactly how time dilation increases in proportion to our spacecraft's speed. Next, we'll look at what happens when it reaches the speed of light.

## 1. The Lorentz factor

The factor by which the clocks are out is called the Lorentz factor, named after Hendrik Lorentz who mathematically described how this works. But what should happen as our imagined spacecraft approaches the speed of light itself? We can make a graph of how the Lorentz factor depends on the speed of the craft; this is what's plotted on the figure on the right. Along the horizontal axis is speed, so the faster you get, the more out of line your two clocks are, which is plotted on the vertical axis. As you go faster and faster, this difference increases almost exponentially. It's not actually an exponential, but it's close in shape. The closer you get to the speed of light, the more dramatic the discrepancy between the traveler and the Earthbound person becomes. In other words, the closer to the speed of light you travel, the longer and longer the Earthbound observer will wait to see one of your seconds tick by.

▼ **Time and motion**
This graph shows the relative slowing of a clock in motion, as a function of how fast the clock is moving. The closer to the speed of light the clock is moving, the slower it appears to tick relative to a stationary clock.

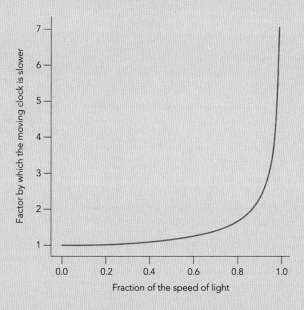

174

## 2. Hitting the speed of light

At the speed of light exactly, time dilation gets a little weirder. The exact relation for the magnitude of the difference between the fast traveler and someone back home on Earth is determined by the ratio of one over the square root of the following:

**(1 − [velocity² / speed of light²])**

If the velocity is equal to the speed of light, then the second part of the equation, in square brackets, is equal to 1. Now subtract 1 from 1, and you get zero. The square root of zero is still zero, which presents us with a problem, mathematically, because we have to divide 1 by zero. Any number divided by zero is infinity. This is why the Lorentz factor suddenly shoots off the top of the figure when you reach the speed of light.

### Verdict

According to our current understanding of the physics, a person on Earth would have to wait an infinite amount of time to observe any amount of time passing for a traveler moving at the speed of light. As far as the Earthbound observer is concerned, time has stopped for the traveler at the speed of light. This doesn't mean that photons—which also travel at the speed of light—are truly instantaneously bashing between objects. Light still has a finite speed, and that's a real measurement of the physics of the universe. Our determinations of light-years are still good, impartial measuring sticks for the universe. But the relative perception of the passage of time means that a traveler going at the speed of light would be able to watch the universe around them age tremendously with every stop that she come to.

▶ After a year on Earth, a spacecraft traveling at different relativistic speeds will appear to have had a different length of time elapse on board. At 10 per cent of the speed of light, the difference is two days; at 95 per cent of the speed of light, only 30 per cent of Earth's time will have elapsed on board the craft.

## DON'T TRY THIS AT HOME

One of the reasons this particular exercise is restricted to the realm of thought experiments is that it would be physically impossible to accelerate a craft to the speed of light itself. In practical terms, speeding up a craft to even one tenth the speed of light is an extremely difficult task. The more you need to speed up your craft, the more fuel you need to bring aboard, which makes your craft heavier, which makes it harder to speed up. In addition, the faster your craft is already going, the harder it is to give it another boost; the amount of fuel you need to increase your speed from 0 to 0.1 c is far less than the amount of fuel you need to go from 0.1 c to 0.2 c. So if you're trying to make it all the way to the speed of light, you'd need a near-infinite fuel source, which makes your craft near infinitely massive, and near infinitely impossible to speed up.

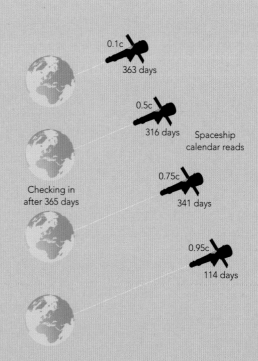

0.1c
363 days

0.5c
316 days

Spaceship calendar reads

Checking in after 365 days

0.75c
341 days

0.95c
114 days

# 6

# GALAXIES

It's time to look outward once more, to the galaxies that host all the stars and their remains. Our home galaxy, the Milky Way, is our portal into the vast array of galaxies that stretch throughout the universe. With a broad range of shapes, sizes, and colors, it is a diverse group indeed.

# WHAT IS A GALAXY?

If you look up at night, far away from the light produced by cities, you can see that our stellar family isn't scattered randomly across the whole sky. There is a clustering of stars, so dense that it might at first appear to be a wispy, high-flung cloud, which stretches across the entire sky in a band. This band is the next step up in our cosmic family tree—it's our home galaxy.

Taking the family tree approach, this galaxy, the Milky Way, is our great-grandparent. Within the Milky Way, all of our grandparent stars are held together, bound by their mutual gravity and the gravity of diffuse gas and dust—the marker of other stellar grandparents that have already come and gone, and the creator of new, future generations of stars. In the early universe, it was the coming together of gas, dust, and dark matter that drew in ever more material, fostering the growth of a gigantic stellar family within each galaxy's gravitational bounds. As the star-forming gas collapses downward into a disk of material, spinning more rapidly than it had been before that collapse, so galaxies are rotating disks of material—on a colossal scale.

A galaxy is a collection of gas, stars, dust, dark matter, and at least one large black hole, which sits at its center, and our Milky Way is no exception. To try to place our galaxy among its sibling galaxies, we examine its size, shape, and color.

## DARK MATTER

Dark matter is a still mysterious component of our universe. It seems to make up about a quarter of all the material in the universe, but it interacts with the matter of stars and planets only through gravity. Many searches for the particle responsible for dark matter are ongoing!

## ELLIPTICALS AND SPIRALS

Unlike stars, which we divided up primarily based on their color, a galaxy's first division is based on shape. The vast majority of galaxies come in one of two shapes: like a slightly deflated ball—roundish along one direction, a little longer in the other direction and fuzzy all over; or like a pancake—an extremely flat, round object, with spiraling concentrations of stars winding outward from their centers. The fuzzy, deflated galaxies are called ellipticals, a term meant to describe their oval shape. The flat galaxies are called spirals, in turn meant to describe their swirling patterns.

Edwin Hubble was one of the first people to try to classify galaxies, and he developed a system that we've since dubbed the "tuning fork diagram" (see page 180), which puts the fuzzy ellipticals on one side (usually the left). The elliptical galaxies are mostly families of very old stars, and are thought to have formed early on in the universe. We think that they might also have been flat spiral galaxies at an earlier point, but that over time, the stars within that galaxy got pushed into more and more random orbits, gradually making them rounder.

▶ The Milky Way, seen from Earth, is a unique vista into the structure of a spiral galaxy, as it is the only galaxy we see from the inside.

Spiral galaxies go on the other (usually right-hand) side of the diagram. Their structure is a bit more complex than the fuzzy, amorphous ellipticals. The flat portion of the galaxy is called the disk, with a central, fuzzier (and less flat) portion at the very middle, dubbed the bulge.

The two tines of the Hubble "fork" diagram mark the next tier of our sorting of galaxies. There exists a set of spirals that have a rather odd structure in their centers—what appears to be a straight line crossing the very center of the galaxy. These structures are called bars. They can be such a dramatic feature of a galaxy's shape that we separate those galaxies that have a bar from those without. Since these are all galaxies with the same general shape (flattened, with winding spiral "arms" radiating outward from their center), they retain the name "spiral" galaxy, but to distinguish them they're called "barred spirals."

It's common for barred galaxies to be drawn with only two arms, but this isn't always the case. We have determined that our own galaxy has a relatively strong central bar, and many more than two spiral arms. In fact, if you look at our best maps of the Milky Way, not only do we have more than two spiral

arms, but we also have smaller branches (sometimes labeled "spurs") between the arms, as the arms themselves fork and divide. It's within one of these little offshoots that our solar system—with the Sun, Earth and every human—resides, about two-thirds of the way out from the center of the galaxy.

There are also exceptions to this general set of elliptical/spiral/barred spiral galaxies, because nothing in the universe is wholly straightforward. There's a class of galaxies that are technically called "flocculent spirals." These galaxies are still pretty flat—they're more like a spiral galaxy than an elliptical—but they're remarkable because they completely lack the spiraling arms. They have tufts and wispy bits, but nothing so dramatic as a spiral arm. ("Flocculent," in case you are curious,

**▼ The types of galaxy**
Hubble's Tuning Fork diagram places elliptical galaxies on the left and spirals on the right, with the latter split into barred and unbarred. Ellipticals are sorted by how round they appear, and spirals by how tightly wound their spiral arms are. Galaxies that don't fit on the diagram are placed in the catch-all "irregular" category off to the side.

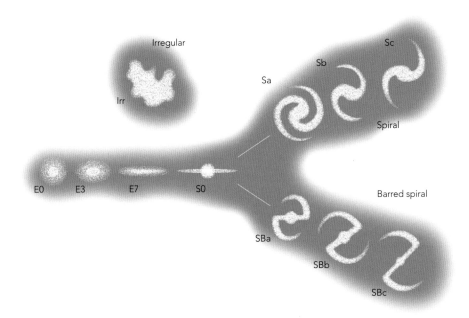

means "wool-like," which is a rather pleasing mental image for describing the disk of a galaxy.) Our nearest neighbor, the Andromeda galaxy, is not quite a flocculent spiral, but nonetheless doesn't appear to have strong spiral arms, showing dark dust lanes instead.

## DENSITY WAVE THEORY

Understanding spiral arms is complicated by the fact that the components of a galaxy rotate at different speeds; the inner part of the galaxy rotates faster than the outer parts. So the easiest explanation—that the arms are actually just areas of the galaxy that are physically more dense, and a fixed component—doesn't work. This is excellently demonstrated by the Sun's magnetic field, which gets increasingly tangled over time, as we've seen. The inner part of the galaxy will rotate many more times than the outer part will, so if the arms were physically planted, like the magnetic field is in the surface of the Sun, we would expect the bright arms to wind themselves into an incredibly tight spiral.

Any solution to this problem has to mean that the spiral arms can't be attached to anything, and they can't be a physical object. So far the best explanation is called Density Wave Theory, which is effectively what happens if you have a traffic jam going in circles. The spiral arm is a density wave, a compression of the material that already exists in the galaxy, but stars aren't fixed to a location "inside" or "outside" of the spiral arm. Like cars passing through a traffic jam, while the stars are within the arm, they're in a very tightly packed region of space (and therefore quite bright, as the light from a large number of stars adds together before streaming toward us), but they're not stuck there for ever, and eventually they will pass onward, as the density wave moves past them.

▼ NGC 1300 is a classic barred spiral galaxy, about 69 million light-years from Earth. Seen here with the Hubble Space Telescope, the central bar is much yellower in color than the blue spiral arms, which begin at the ends of the bar.

# HOW MANY GALAXIES ARE THERE?

Just as the Sun is part of an astoundingly large stellar family, so our galaxy is also part of a universe-wide family of galaxies, separated by huge volumes of space. If we look around our home galaxy, it can be difficult to appreciate just how many other galaxies are out there. However, if we look beyond the Milky Way, we can capture a glimpse of how big this family is.

The Hubble Ultra Deep Field is one such image of the distant universe, and it gives us a flavor of just how galaxy-rich our universe really is. I encourage you to find a high-resolution version online and look around. Every single bright pixel in this image is a galaxy (with the exceptions of the bright spots with spikes—those are stars in our galaxy that got in the way).

The patch of sky imaged for the Ultra Deep Field was chosen because it is particularly dark, thus avoiding light contamination from our own galaxy. Certain parts of the sky are totally unusable for studies of distant galaxies; the gas and dust in our own galaxy is simply too bright in those directions. You would also struggle to find anything directly behind a nearby galaxy. You really want to be looking at the blackest, emptiest part of the night sky you can find. The down side to choosing an extremely dark patch of sky is that the galaxies you're interested in seeing are very faint. Even the best telescopes have to dedicate an extraordinary amount of time to capturing the light from these distant galaxies.

For the Ultra Deep Field, the Hubble Space Telescope spent nearly 1 million seconds staring at this patch of sky. If this had been observed in a single session, it would have lasted for a little over 11.5 days. If you could spend the same amount of time on every patch of sky, you would expect to receive a similar-looking image every time. This means we can take the Ultra Deep Field, with its 10,000 galaxies, and imagine tiling it out across the sky. Given how small this image is on the

## 12 MILLION GALAXIES

The Ultra Deep Field is so small on the sky that 1,200 of these images would fit under your little fingertip. Face in any direction, look up at the night sky, and raise your smallest fingertip—you've just blocked out the light from over 12 million galaxies.

sky (see the box above), such a tiling would be a very fine mosaic indeed. Our galactic family is not small; with 12 million galaxies per little finger at arm's reach, we rapidly reach billions of galaxies before we've covered a large fraction of the sky.

▶ This image is a composite of separate exposures taken from 2003 to 2004 with Hubble's Advanced Camera for Surveys and Wide Field Camera 3. It assembles one of the most dramatic deep space images ever captured by the Hubble Space Telescope, and it offers us a comprehensive picture of the evolving universe as well as a humbling sense of our place in the universal family tree.

ASTROQUIZZICAL

182

# HOW CAN WE COUNT THE STARS IN OUR GALAXY?

Most of the stars in our night sky are part of the family that make up our own galaxy. While it is relatively straightforward to chart the locations of the brightest stars, getting a more complete census of exactly how big our family of stars is, is a much more difficult task. Current estimates put the number of stars in the Milky Way at somewhere between 100 and 400 billion, which is not particularly precise.

**E**ven by astrophysical standards, a factor of four, changing our count by 300 billion stars, is doing particularly badly. We can actually do considerably better for other, much more distant, galaxies in our local universe. A factor of four is me telling you that your dog weighs somewhere between 60 and 240 pounds (27 and 109 kilograms). A 60-pound dog may be medium to large, but a 240-pound dog is a different creature entirely (it's a bear). So why can't we do much better? It's because we're sitting within it. We can't get a good viewing angle, to see what shape our galaxy really is, or where the stars are located.

So what can we do? A very basic path would be simply to try to count all the stars we can see, and see how far we can get. Some of the original maps of the galaxy were made this way, but by the time our estimates get all the way up to 100 billion stars, there's no way we're going to have enough time to count them. If we're right that there are at least 100 billion stars, and you only spend a second looking at each one, you're stuck with at least 3,180 years' worth of work laid out in front of you. That's not insurmountable, if you divide and delegate well enough, but there's another, more severe problem with the counting method.

The galaxy is full of gas and dust, both of which block out the light from more distant stars. Clouds of gas and dust tend to be the densest, and therefore the most opaque to starlight, within the disk of the Milky Way. Within the disk is exactly where our

solar system is, along with most of the stars in our galaxy. So not only are we most prone to missing stars as we look along the disk of the galaxy, but we're also missing the largest number of stars! Simply counting the visible stars would vastly underestimate the number of stars in our cosmic family.

## INFRARED VISION

There's a slightly more sophisticated method we can use to try to get around this "dust blocks light" problem. In the optical regime, which is where our eyeballs function, dust will always obscure light coming from behind the dust cloud. If, however, we look in the infrared, the dust is transparent, and we're able to catch the glow of stars from both within the dust cloud, and beyond it. So if we can measure the total amount of light coming from our own galaxy in the optical, plus whatever it's doing in the infrared and other wavelengths, we can partially account for the light that might be missing. The trick with this method is that you now have to figure out how many stars should be responsible for the creation of a fixed amount of light.

To do this backward calculation, you need to know how many of each size star you expect the galaxy to make. If your galaxy is good at forming very large stars, you need fewer stars to produce the same amount of light, because those large stars are also the brightest. If the galaxy is good at making very small stars, you could have a lot more individual stars, because the small stars are the faintest. You can add quite a few faint stars into the galaxy before you change the amount of light produced by a significant amount. Galaxies produce stars of all sizes, but the exact ratio of the number of small stars to the number of large stars is something astrophysicists are still trying to measure. We can make this measurement for the stars very near to us, but the fainter the star, the harder it is to spot. This uncertainty on how many large stars we should expect is partly why the total number of stars in our stellar census is so variable—and it's hopefully something we'll be able to nail down a bit better as research progresses.

▼ The exact same section of the sky, looking toward the center of the Milky Way, seen in optical light (top), where dust obscures much of the starlight, and in the infrared (bottom), where the dust is almost entirely transparent.

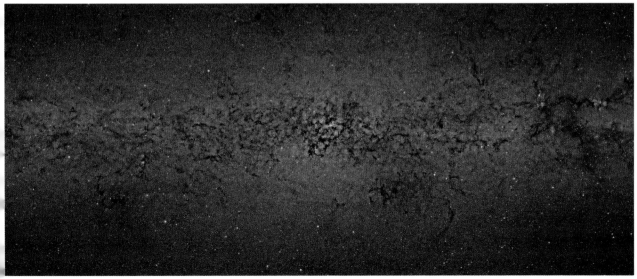

# WHAT IS A SUPERMASSIVE BLACK HOLE?

As far as we can tell, all galaxies contain a supermassive black hole at the very center of their bulge. Unlike the black holes discussed in Chapter 5 (see pages 160–169), which form when massive stars collapse and are a few times larger in mass than the Sun, a supermassive black hole is 1 billion times more massive than the Sun.

All our observations suggest that black holes and their galaxies grow in lockstep with each other; smaller supermassive black holes are found in smaller galaxies, and larger supermassive black holes in larger galaxies. In addition, one supermassive black hole seems to be the rule. While it's possible to spot a galaxy with two gargantuan black holes at its core, this kind of scenario is almost exclusively encountered if the rest of the galaxy appears to be particularly unusual. The assumption in these cases is that the double black hole galaxy has stolen its second black hole from an unlucky nearby galaxy that wandered too close and was devoured entirely. The unusual shape of the rest of the galaxy would then be the hallmark of the chaotic aftermath of having consumed that galaxy.

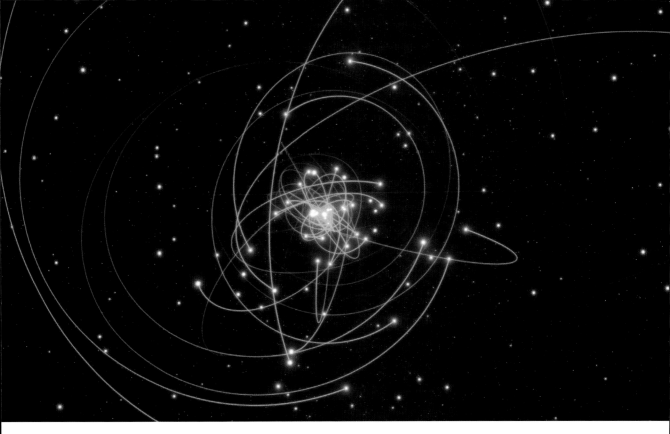

▲ The orbits of stars very close to the supermassive black hole at the heart of the Milky Way, known as Sagittarius A*. One of the stars that passes closest to the black hole is dubbed S2, and it orbits the invisible black hole once every 16 years.

◀ The Hercules A galaxy, as seen by Hubble, is merely the yellow-white fuzzball at the center. However, the Very Large Array sees huge jets, many times larger than the galaxy itself, driven by the galaxy's supermassive black hole. These jets glow brightly in radio wavelengths (shown in red).

It's worth reiterating here that the black hole at the center of our galaxy isn't doing much of anything, other than sitting there and being massive. Much like our solar system is in a stable orbit around the Sun, the vast majority of a galaxy is in a stable orbit around its center, with no real reason to go plunging toward the black hole at its very center.

In fact, the region around a black hole can be quite barren. The Milky Way's central black hole, for instance, seems to be surrounded by stars but almost no gas, so there's nothing actively swirling around it in a heated disk. In order to be shredded by a black hole, a star would have to come extremely close to it. Some of the stars that orbit the black hole in the center of the Milky Way go around it once every 16 years and we've been able (incredibly) to watch them move around it. Some stars come within a light day (16 billion miles, or 173 astronomical units) of the event horizon, and that's still not close enough to get torn apart or sucked in. (There are videos of the orbiting stars. Watch them; they are very cool.)

# IF THE UNIVERSE IS EXPANDING, HOW CAN TWO GALAXIES COLLIDE?

Gravitationally driven collisions between galaxies aren't particularly common in our universe, but they're not especially rare either. This may seem surprising, given that the universe is expanding. The apparent contradiction stems from a set of simplifications commonly used to explain the expansion of the universe.

One such simplification goes like this: "Imagine a balloon with a series of dots on the outside of it. Now inflate the balloon. All the dots move away from each other as the balloon, which is space, grows in size." Another involves bread and raisins: "Imagine you have a loaf of bread with raisins in the surface. As the dough rises, the raisins will spread farther apart from each other."

The fabric of space is indeed expanding as the universe ages, and these metaphors capture the concept of the distance between objects expanding, rather than the objects themselves moving at high speed. However, there's a tendency to illustrate this metaphor—the dots on the surface of the balloon, or raisins in bread dough—with regular patterns. We put everything on a grid, so that the effects of the universe's expansion are easier to spot. To some degree, this is convenient because it means we're only looking at one effect (the expansion), but it's a big oversimplification of the structure of the universe. Objects in the real universe aren't laid out on a grid. The universe doesn't do grids. Real galaxies are scattered more irregularly across the fabric of space, which means that sometimes you're going to wind up with one or two or 50 galaxies pretty close to each other. Sometimes you'll wind up with a galaxy with nothing around it at all.

When you have two enormously massive objects relatively close in the universe, the force of gravity takes over. If two galaxies are near enough to feel the gravitational pull of each other, it doesn't matter that the universe is expanding—it isn't expanding fast enough to counteract the attractive force of gravity, and these two objects will inevitably fall toward each other. When they do, there's a good chance they will eventually become a single, larger galaxy, and the process gives us magnificent images.

Real galaxies are scattered more irregularly across the fabric of space, which means that you're going to wind up with one or two or 50 galaxies pretty close to each other.

Interactions between galaxies can cause a galaxy to be strongly distorted, due to intense gravitational tides. Like the tides on Earth, these stretch the galaxy out, and pull material out into two "arms." Because these arms are caused by an external influence in the form of tidal forces, they are given the name "tidal arms." These tidal arms are often quite dramatic, stretching across many hundreds of thousands of light-years. They are markers of a recent swing-by of the two galaxies, and often foreshadow a more dramatic collision of the gravitational companions in the relatively near future, when they will merge into a single object.

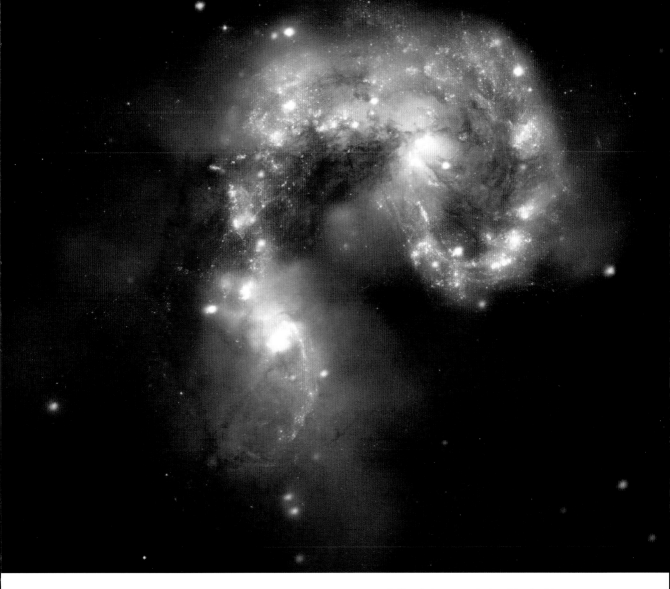

What of their black holes? After things settle down, the heaviest objects will end up at the center, which for two galaxies is their two supermassive black holes. Over time, the two black holes will lose enough energy while orbiting each other to merge into a single black hole, as we saw earlier. If the merging galaxy were about the same mass as the original galaxy, this should double the mass of the central black hole in one fell swoop—a much more efficient path to growing the size of a black hole than trying to build mass with thin streamers of gas.

▲ The Antennae galaxies shine in this multiwavelength composite. A Hubble image provides the gold and brown colors, infrared from the Spitzer space telescope is shown in red, and X-ray light is in blue. These two galaxies have been tangled in a galactic collision for at least 100 million years, with many more millions of years to go before they become a single galaxy. These two galaxies are currently forming a prodigious number of new stars, which contributes to the infrared glow.

# WHAT WILL BECOME OF THE MILKY WAY?

With all this talk of galaxies that are too close to each other eventually merging, and given how close we are to our neighboring, slightly larger galactic sibling of Andromeda (about 2.5 million light-years away), we might well wonder what influence Andromeda has on the Milky Way's future.

The Milky Way is accompanied not just by the Andromeda galaxy, but by the entire Local Group, which contains some 50 galaxies. But Andromeda is by far the most significant of these, weighing in somewhere between 700 billion and 1 trillion solar masses. This is approximately the same mass as the Milky Way, which is also usually considered to have about 1 trillion solar masses' worth of stuff hanging around. The rest of the Local Group are mostly not very massive objects (like the Large and Small Magellanic Clouds), which are gravitationally tied to either the Milky Way or Andromeda, and which orbit the larger galaxy to which they're bound.

The Milky Way and Andromeda, as the largest objects in the Local Group, are also gravitationally tied to each other, and orbit each other extremely slowly. However, to determine where these orbits will take our galaxies in the future, you need to know a bit more about them than just their current positions and their mass. Both of those pieces of information are critical, but we also need to know how fast they're moving relative to each other. With this information, you can determine what path the two objects will take in the future.

The masses of two objects in space determine the point around which both objects will orbit. This is called the center of mass, and as we saw earlier, it is defined as the point in space that has an equal distribution of mass around it. For a system like the Sun and Earth, the Sun contains almost all the mass—Earth is far enough away from the Sun that its relatively tiny mass doesn't alter the center of mass very much. Earth only pulls the Earth–Sun center of mass the tiniest bit away from the center of mass of the Sun itself. If the two objects are similar to each other in mass, however, the center of mass is between the two, in empty space.

This is the case for the Milky Way and Andromeda. They're both extremely massive objects, but neither can escape the gravitational pull of the other. They both orbit around a point somewhere near the middle of the space between them. This point—the center of mass between the two—is the location of an inevitable intergalactic collision in 3–5 billion years' time. Our two galaxies aren't moving fast enough relative to each other to avoid this point where our galaxy and Andromeda will eventually collide. Some time after that, the supermassive black hole that currently sits at the center of our galaxy, and the one at the center of the Andromeda galaxy will merge together, creating an even more supermassive black hole.

It's tempting to think that our Sun and our solar system will be caught up and destroyed in this cosmic collision, but in truth the vast distances between the stars in both galaxies mean that no two stars will hit each other. The Sun's orbit around our newly merged, enlarged galaxy will certainly change, and the stars in the night skies of our planet will jumble, but the solar system will persist intact throughout the long process of the Milky Way's collision with Andromeda.

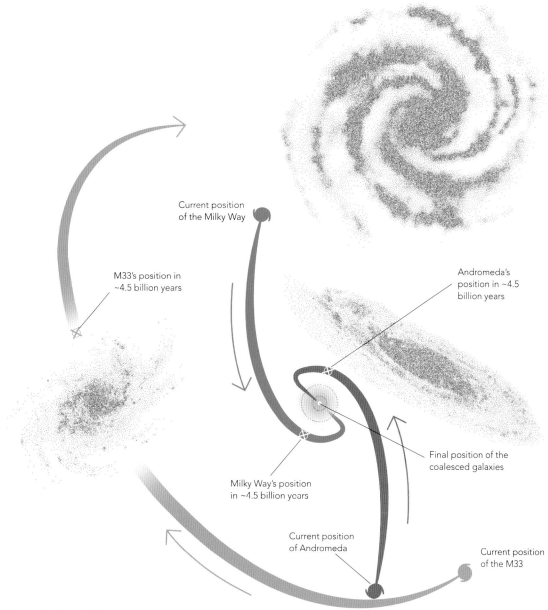

Current position
of the Milky Way

M33's position in
~4.5 billion years

Andromeda's
position in ~4.5
billion years

Final position of the
coalesced galaxies

Milky Way's position
in ~4.5 billion years

Current position
of Andromeda

Current position
of the M33

▲ **When galaxies collide**

The Milky Way (blue) and Andromeda
(red) are forecast to collide with each
other; their future paths are charted out
here. These two galaxies are due for a
remarkably direct and rapid collision,
while the companion to Andromeda, M33,
will mostly miss the gravitational shake-up.
Andromeda and the Milky Way will merge
into a single galaxy about twice as large
as each galaxy is now.

# WHAT IS A QUASAR?

The black holes at the centers of the Milky Way and Andromeda are currently quiet, without any material actively falling toward them. But when a supermassive black hole manages to gather a lot of material to itself, it can light up very dramatically.

Supermassive black holes can produce an absolutely stupendous amount of energy: millions of times the energy output of the Sun. This energy is a by-product of the extraordinary inefficiency with which a black hole manages to absorb external material in order to grow. The material falling inward gains a lot of energy, both in terms of the speed with which it is orbiting, and in heat. If only a tiny fraction of that material is actually absorbed by the black hole, the rest of it has to go somewhere. The energy produced by the black hole often funnels out into a bright jet of material, flung outward along magnetic field lines at speeds that reach significant fractions of the speed of light.

These jets can only appear if the black hole is actively trying to accrete new material, and we might reasonably expect the galaxy it sits in to react to the flinging of these jets. There's so much energy being produced that it seems only natural to think the galaxy should somehow care. This is still an active field of research, and we haven't quite converged to a consensus on how the black hole should shape the behavior of the rest of the galaxy. But a good place to look for such hints of interactions between the black hole's operations and the galaxy as a whole is in a class of galaxies with very extreme black hole behavior: quasars.

## ENERGETIC BLACK HOLES

The term "quasar" is a portmanteau of "quasi-stellar radio source," though nowadays the term has been retroactively turned into a portmanteau of "quasi-stellar object," which then got abbreviated further into QSO. The "radio" part was dropped from the naming convention because it turned out to be a poor description.

Quasars are the observational signatures of a particularly energetic black hole, doing a very good job of flinging material away from a galaxy. To an Earth-based observer, quasars are extremely bright dots of light on the sky, looking very much like a star. Initially, these bright spots were puzzling, because a detailed investigation of their light spectrum revealed that they were like no star we'd ever seen. Because quasars blasted the camera with light, it was very difficult to determine their context—were they located in a specific kind of galaxy, or at a particular location in the galaxy?

It eventually became clear that these quasars are indeed in the centers of galaxies, and the galaxies they're located at are incredibly faint, relative to the brightness of the quasar. It was established that the light from quasars had to be produced by the central black holes of these faint galaxies, which were dramatically outshining the rest of the galaxy. It also turned out that quasars are much more common at large distances from us—so they are much more common in the younger universe.

How did they get so bright? To understand that, we need to poke our metaphorical noses a little closer into the black hole. An active black hole (in that it is trying and mostly failing to accrete new material and grow) is always surrounded by a superheated disk of material—an accretion disk—which is swirling around the black hole, gradually losing the angular momentum it needs to maintain a stable orbit, and slowly falling inward. This innermost accretion disk is often surrounded by a larger donut-shaped ring of dust and gas, which is not quite as hot, and not nearly as close.

▲ This illustration of a quasar shows a gas-rich accretion disk surrounding a black hole. Material falling into the black hole is heated to tremendous temperatures, but to be seen as a quasar, the galaxy must be oriented so that our view of it from Earth is directly down the energetic jet of material spun out of the galaxy by the black hole. These objects are rare, but much easier to spot than another galaxy at their distance from Earth, given how bright they are.

This disk of material will be surrounded by a magnetic field, like the magnetospheres of Earth and the Sun, and so the easiest escape path for a particle is perpendicular to the disk, where the magnetic field can't get as tangled. Particles trying to escape find the easiest way out, which is always "up" or "down" from the disk, so they are funneled rather efficiently into a jet.

However, the particles that get fired out into this jet are mostly electrons and protons, not packets of light (photons). Getting from a beam of electrons to a beam of photons requires a few transfers of energy. One method of energy transfer happens when electrons travel along the untangled magnetic field lines extending outward along the beam. The magnetic field causes the electrons to travel in a helix, as though they were tracing the path of an extremely long spiral staircase. These spiraling electrons give off high-energy particles of light. If the electrons are moving fast enough, they can spit out gamma rays; if they're moving less fast, they can radiate their energy away through X-rays and radio waves.

You can also create high-energy photons (up to gamma rays) with a more straightforward path: via collision. If you take an extremely speedy electron and crash it into a photon, the photon can gain enough energy to become a gamma wave. The end result of both of these processes is a pencil beam of high-energy radiation speeding outward in two jets away from the accretion disk of the supermassive black hole.

GALAXIES

## LOOKING A BLACK HOLE IN THE EYE

If you happened to be pointed in a direction that had you looking straight down the jet, you'd find yourself staring down a very bright portion of the black hole, instead of being able to see the glare from a less direct path. It seems that with quasars, we happen to be looking down the jet, or very nearly directly down it. The jets produced by quasars are very stable, and can reach lengths of hundreds of thousands of light-years.

Orientation can't be the only key to spotting a quasar, because then we'd expect to see them nearby as well, and we don't. So there must be a secondary factor at work as well; very likely this is the amount of gas hanging around near the center of the galaxy. If there's not enough gas in the galaxy to create that innermost disk, the black hole simply won't have any material to fling outward, much like the situation in the Milky Way. In the local universe, galaxies tend to be mostly stars, with less than 25 per cent gas. But galaxies that existed at earlier times in the universe have the percentages reversed: they can easily be more than 60 per cent gas, with fewer stars. This means that even without any external effects from things like collisions between galaxies, it's much easier for the black hole to have an easy fuel source to power its huge jets, streaming away from the galaxy.

▶ The massive elliptical galaxy M87 is seen here by the Very Large Telescope in Chile. The jet driven by the supermassive black hole at the center of this galaxy is visible in optical light. M87 is one of the most massive galaxies in the nearby universe, and so it contains one of the largest supermassive black holes, able to drive extremely strong jets. It is this galaxy's black hole that was imaged by the Event Horizon Telescope in 2019.

▼ **Synchrotron radiation**
When a charged particle, such as an electron, moves through a region with strong magnetic fields, it will spiral along the direction of the magnetic field. The electrons then give off light, known as synchrotron radiation. Depending on the strength of the magnetic field and the speed of the electron, this radiation can be of different wavelengths.

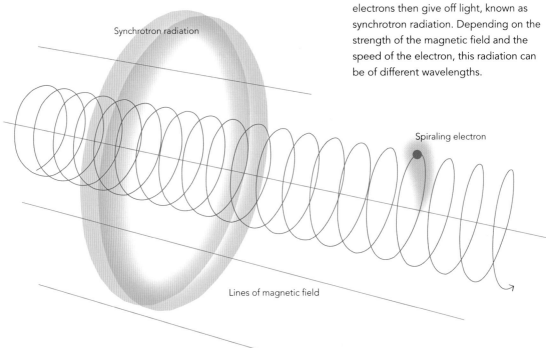

Synchrotron radiation

Spiraling electron

Lines of magnetic field

# WHAT IS GRAVITATIONAL LENSING?

Space is distorted around massive objects, and the more massive the object, the more extreme the distortion becomes. Since galaxies are fairly hefty objects, they can create substantial warps in the space surrounding them. The easiest way to notice? Light will travel in a straight line, but a straight line through curved space is a curve. The presence of their mass bends the path of light.

The degree to which light bends around a galaxy is not as extreme as around a black hole, because the distortion of space surrounding an entire galaxy is less steep. A galaxy is not nearly as densely packed as a black hole.

Similar to the way that a magnifying lens changes the path of light and magnifies it, the gravitational distortion surrounding a galaxy can also function as a lens. Indeed, this bending of light is technically called gravitational lensing. If you have a look at the objects along the edge of a photo taken with a fisheye lens, you'll notice that things that would be straight lines (like walls or trees) are bent into a curve—the light traveling through the lens was distorted as it passed through. The fun thing about

a gravitational lens is that it means we can spot objects that would have been too distant or too faint to see without the magnification that the gravitational lens was able to provide.

The gravitational heavyweight that serves as the distorter, bending the light from the object behind it, could be a single, massive galaxy. (Equally, a cluster of galaxies, which collectively distort the space around them, can do the trick.) In order for us on Earth to see the lensing effect, the alignment of Earth, the heavy galaxy, and the distant object has to be exactly right. We want the light from the distant object to be almost exactly lined up, precisely behind the galaxy, from our viewpoint. The closer to a perfect alignment you can get, the

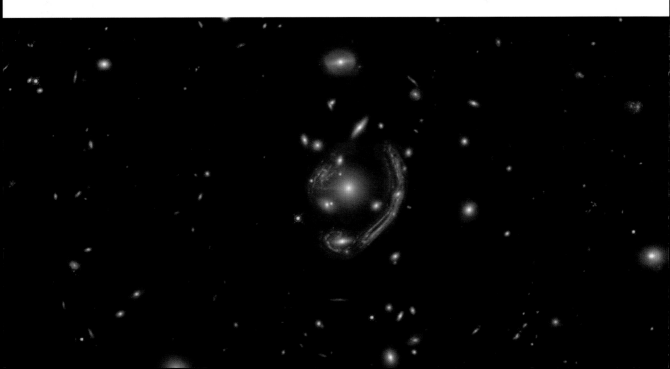

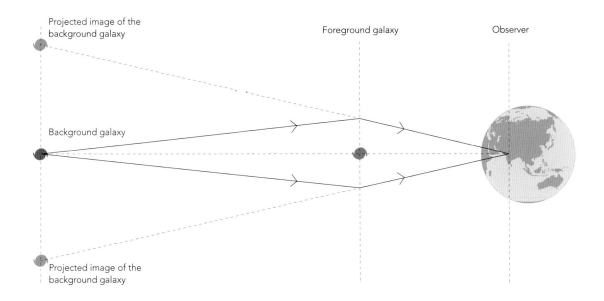

Projected image of the background galaxy

Foreground galaxy

Observer

Background galaxy

Projected image of the background galaxy

▲ **Gravitational distortion**

A bird's-eye view of the light path from a background galaxy, which travels out in all directions. When some of the light encounters the gravitational distortion of a foreground galaxy or cluster, it is bent so that it arrives to the viewer on Earth. We then see the galaxy as though it appears multiple times in the sky.

◄ A background galaxy is dramatically distorted around a foreground elliptical galaxy in the Fornax cluster. The galaxy, geometrically, is almost perfectly behind the central elliptical galaxy; light has found many paths around the gravitational distortion caused by the cluster on its way to our viewpoint on Earth.

more distorted the light will be, as it tries to go through the most affected parts of the space surrounding the galaxy. If everything is precisely lined up, then the light from the background object will pass around the heavy object in front of it in an even way, like water flowing over a sphere. As we see it, there would be a perfect ring of light from the background object around the heavy front galaxy (or galaxy cluster). This is what's called an Einstein ring, because it's a perfect demonstration of the prediction of light's path made by Einstein.

However, most of the time the background object, the heavy lensing object, and we as observers are very slightly out of line. This means that light isn't passing around the heavy lensing object in an even way. More light will bend around one side, leaving an empty space on the other side. If the background object is offset enough, you wind up with multiple distorted images from that object—they'll just look a bit stretched, like the fisheye lens effect. This is the effect that caused the "Space Invader galaxy."

Usually, the background object behind the heavyweight is another older, more distant galaxy, and it's the light from this galaxy that is sheared out and stretched like silly putty. The foreground gravitational weight really just has to stretch space for this to work; but if you want a perfect Einstein ring, you have to cross your fingers and hope for a cosmic alignment.

# Haven't the stars in distant galaxies died out by now?

As we look out into our universe the most distant objects we can see are also the most out of date; their light is quite old by the time it reaches Earth. If we know a few things about how stars age and cease to shine, it's a sensible thing to wonder whether any of the stars we see are still around. How do distant galaxies *actually* look now?

The field of galaxy evolution is a game of trying to solve exactly this question. Given a set of snapshots of galaxies at different points in time, which types of galaxies are likely to transform themselves into which other types? To understand these changes, the biggest piece of the puzzle is understanding what happens to the stars within the distant galaxy. We'll examine a couple of competing pathways and then set our mind to a fascinating conundrum relating to the gas every galaxy contains.

### 1. Do stars die out or are they replaced?
If we're looking billions of years into the past, a large number of the stars that shine in our dated images of them will have either exploded into supernovae (if they are massive enough), or transformed into a red giant and left behind their white dwarf remnants. The smallest and least luminous of the stars in these images will remain, churning out their reddish hue into the space surrounding them. We have found extremely old stars in our own Milky Way, so we know that these stars shouldn't go anywhere. The real question that a teleportation would answer is: Did those bright blue stars get replaced or simply die out?

If galaxies formed their stars all at once, early on, and then did nothing more, we should expect to see them change fairly dramatically with time. The bright blue stars that are the most luminous in any galaxy will die off relatively quickly, outlived by their dimmer, redder siblings. In the absence of any other external forces we'd expect to see the galaxies more or less in the same shape they were in the past, but just extremely red, rather than blue.

If we look in the nearby (and therefore relatively recent) universe, however, we see that plenty of galaxies are still forming new stars, so this "burst of star formation and then nothing" idea is probably not correct. On the one hand, this is reassuring, because we also don't see red copies of the lumpy, oddly shaped prototypes of spiral galaxies hanging around in the local universe, and it's nice to have a reason to point to for why that might be. On the other hand, this makes our life more complex, because now we have to think about a galaxy's budget of gas against the rate at which it forms stars.

If a galaxy gets no additional gas, it will continue to form stars until it runs out of gas. If, however, a galaxy gets refueled, it can continue to form new stars practically indefinitely, as long as the refueling process keeps up. If a galaxy can continue to form new stars, it will likely remain relatively blue with the light of the brightest and youngest stars.

### 2. How do you refuel a galaxy?
The easiest way to get more gas into a galaxy is to throw another galaxy with some gas at it. Since we know that galaxy interactions and collisions are not that uncommon in the universe, this is a relatively plausible explanation. But because not all galaxies live near other galaxies, some of them will be refueled more frequently than others.

This means that—depending on the galaxy, how much gas it currently has to maintain the formation of stars, how many other galaxies are nearby to continue fueling it into the future and whether that other galaxy has enough gas to effectively refill the galaxy's gas—you could get a wide range of galactic appearances in a span of several billion years!

## Verdict

It's really hard to say exactly how a specific galaxy—say one in the Hubble Ultra Deep Field image—will change over time, as it depends so strongly on factors outside the galaxy and how they change with time. But with any of these galactic outcomes, we know it won't look the same.

## EVOLVING GALAXIES

- A galaxy that merges frequently, but with galaxies without much gas, is likely to transform itself into a massive elliptical galaxy.
- A galaxy that lives in an empty region of space, with no real means to refuel and no galaxies nearby to disrupt it, is likely to simply fade to red.
- A galaxy that can encounter other gas-rich galaxies every so often is likely to continue to produce bright, blue stars, while gathering mass; galaxies like the Milky Way or Andromeda were probably built this way.

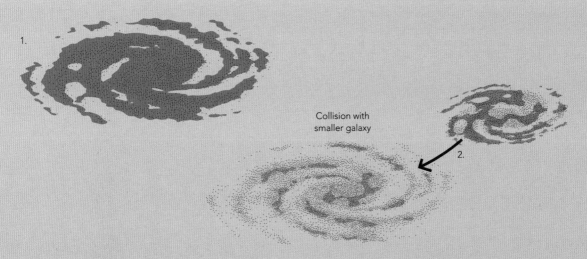

1.

Collision with smaller galaxy

2.

Refuels: stars keep forming

3.

▲ **Galactic refueling**
A galaxy that begins with a large reservoir of gas will eventually consume that gas as it converts it into stars. Without the intervention of another supply of gas (such as a collision with another gas-rich galaxy), it will eventually exhaust its ability to make new stars. Collisions, by contrast, can "refuel" the galaxy's reservoir of gas, allowing it to continue forming stars.

GALAXIES

# THE UNIVERSE

The galaxies strewn around the universe are diverse: some are merging, and others are forming stars; some are located within a cluster of thousands of other galaxies, and others are relatively isolated. By studying the way galaxies are distributed, we learn about the next tier of our cosmic family tree: the fabric and structure of the universe itself

# WHERE DID THE GALAXIES COME FROM?

We've traced our way up from parent planet Earth, to grandparent Sun, to great-grandparent Milky Way, and now it's time to rise another step higher in our cosmic family tree to our cosmic great-great-grandparent: the universe itself. But what is left in the universe, which provided the means for galaxies to form?

The universe, as far back as we can measure it, is remarkably similar in all directions. It's not perfectly similar in all directions—otherwise we wouldn't expect to see individual galaxies, separated by wide distances. What "remarkably similar" means is that the number of galaxies in any given direction seems about the same, and the galaxies that all have a particular shape aren't all over to the left and a different shape over to the right—the types of galaxies are also pretty evenly spread. So where did these galaxies come from?

▼ **A timeline of the evolution of the universe**
The oldest light we can observe directly was emitted when the universe was 380,000 years old; this was the moment when the universe became transparent to light. Small changes in density, set in place randomly in the early universe, grow into dense regions in the later universe, becoming the cradles for galaxies and clusters of galaxies to form, eventually constructing the universe we see today.

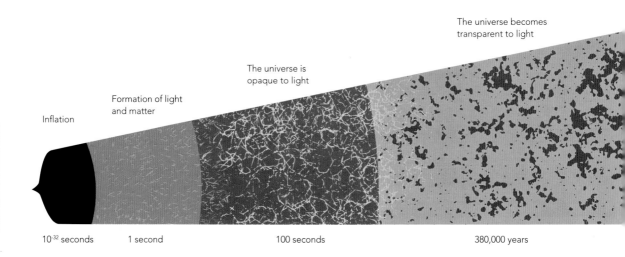

The universe becomes
transparent to light

The universe is
opaque to light

Formation of light
and matter

Inflation

| $10^{-32}$ seconds | 1 second | 100 seconds | 380,000 years |

In the very earliest light we can detect within our universe, we observe a cloud of hot, dense, high-energy particles, currently measured at a fantastically even temperature of 2.725 Kelvin (see the box on the right) in every direction we look. This is reassuring, because if this cloud were very uneven, we'd expect that unevenness to be reflected in our current universe. However, if we look carefully, with precise enough instruments, we can measure very small temperature changes across the sky. These temperature wiggles are a reflection of tiny density fluctuations in the earliest universe. These tiny fluctuations are so small that we don't need anything fancy to explain them. Purely random collections or sparsities of particles are enough to explain where these wiggles come from. Once a density wiggle came into being, it collected more material to itself, growing larger until that smallest of overly dense regions became the seed for the galaxies and all their stars.

## THE KELVIN SCALE
2.725 Kelvin is the equivalent of –454.7°F, or –270.4°C. Kelvin is a temperature scale like Celsius, but instead of having zero as the freezing point of water (32°F, or 0°C), zero Kelvin is an absolute minimum temperature. Absolute Zero is defined as zero Kelvin; 32°F corresponds to 273K.

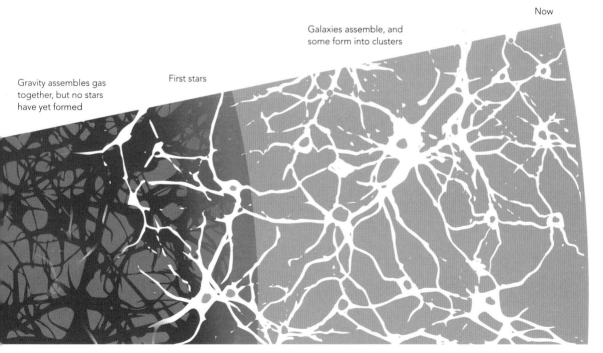

Now

Galaxies assemble, and some form into clusters

First stars

Gravity assembles gas together, but no stars have yet formed

300–500 million years

Billions of years

13.8 billion years

# WHY IS IT SO DIFFICULT TO OBSERVE THE UNIVERSE?

In trying to understand planets, stars, and galaxies, we've always taken the approach of looking at as many of them as we can, and then using the information about how frequently different styles of cosmic object have appeared to learn something about how they must have formed. But we only have one universe to observe, and much like the problem of trying to figure out the size of our own galaxy, we're stuck inside that universe. This makes observing it is tricky, because our own position determines which part of the universe we can actually spot.

The biggest limiting factor is time. We can only record the portion of the universe surrounding us, where there has been enough time for light to leave an object, travel across space and actually reach human observers on Earth. As a cosmic speed limit goes, the speed of light is high, but considering the distances involved between galaxies, it starts to feel rather constraining. With a start date on the universe of 13.7 billion years

▼ The center of the galaxy (~26,000 light-years distant) is close compared to the scale of the entire galaxy. Light leaving a star on one side of the Milky Way would take more than 100,000 years to traverse the whole galaxy.

ago, the most distant things visible to us have been sending light in our direction for roughly 13.7 billion years. But of course we also see those things as they were 13 billion years ago, not as they might appear if we could teleport to them now.

In the nearby universe, light's limitations are not a major constraint, as the time delays they impose are relatively manageable. The delays even to communicate with the *New Horizons* spacecraft, at the very outskirts of our solar system, are only a few hours (for light, it's a 9-hour round trip from Earth to Pluto). From Earth to Mars is a positively rapid 14-minute round trip, on average. Within our galaxy, it starts to become a little more noticeable that reports of events are delayed in getting to us; for instance, our current observations of the star Eta Carinae indicate that it is probably going to explode sometime soon. However, light takes 7,500 years to reach us, so if it had exploded last week, we'd have to wait that long before we'd be able to notice.

As a cosmic speed limit goes, the speed of light is high, but considering the distances involved between galaxies, it starts to feel rather constraining.

The farther out we go, the more dramatically our vision lags behind "now." This lag affects light equally as it comes from all directions, so we end up with spheres of lagged space-time all around us. Because we are at the receiving end of all this light, this observable sphere is by definition centered on us. But, as we noted earlier, any other observer at any other position in the universe should see the same thing—their view of the universe will be just as limited in scope by the speed of light as our own.

## THE OBSERVABLE UNIVERSE

The section of the universe that we can actually receive information from is known as the observable universe, and that's all we have to work with in deciphering how the whole thing works. This is not a catastrophic problem, because anywhere we look in our section of the universe, physics seems to work in the same way. Extrapolating outward to say that physics should work in the same way even in the parts we can't observe is not an outrageous stretch of logic. Of course, we have no idea how much universe we're missing—or even if there's a number we could put to how much we're missing. Since we live when we do, not enough time has passed since the Big Bang for us to get information from more distant reaches of the universe.

The universe beyond our view may well be infinite. Sometimes saying that the space in our universe is infinite makes the mathematics of describing it a little easier to manage. However, we don't have an infinite amount of time for observation, in order to test the precise infinite or finite nature of the expanses of space, so we won't ever be able to determine the exact size of the entire universe. Even with just our observable portion, though, we can predict a number of things about how the universe at large should behave, and test those predictions against other observations of our little bubble. It's through this method that we've untangled how the universe has evolved over time, which gives us information about how it began, and how it will unfold in the future. We cannot observe the entire observable universe in one fell swoop, however, much as some astronomers (myself included) would love that. In practice, observations are undertaken through a number of paths, some more patchwork than others.

Each telescope facility gives a unique view out onto the universe, tailored to different scientific questions. Some may be optimized for hours-long exposures of small patches of sky, in order to capture the faintest traces of light from a distant planet or galaxy. These tiny patches of sky are chosen by individual scientists, and while most of the time there are checks in place to make sure that four people aren't repeating observations of the same area of sky, there's a lot of sky, and it would take a very long time to cover all of it.

The next strategy is to cover a slightly larger area of the sky, but not to sit on each part for quite as long. In general these fields are about the size of the full Moon on the sky, and many of them are named. Among the most famous of these is the Hubble Deep Field (along with the similar, but deeper, Ultra Deep Field, and an even deeper, smaller subset of the Ultra Deep Field, the eXtreme Deep Field), though both deep fields are about 10 times smaller than the full Moon.

There's one more way to observe the universe— an all-sky survey. All-sky surveys try to look at the entire night sky. For telescopes on Earth, this is limited by where they're planted—a telescope in the northern hemisphere can't observe something in the southern hemisphere's sky. If you put your telescope in space, you can avoid this problem. Once you're there, depending on what wavelength of light your detectors are sensitive to, each telescope will return its own unique census of what's out there. These all-sky surveys range from Planck, which sent back the most detailed image of the universe's oldest light, to Gaia, which returned the very precise locations of the stars in the Milky Way. The edge-on image of the Milky Way shown on pages 204–205 was generated as part of an all-sky survey taken by Gaia.

▶ The galaxy cluster Abell 2744, as seen by the Hubble Space Telescope, is one of the deepest images of any galaxy cluster. These cluster galaxies are 3.5 billion light-years distant from Earth, but this image also includes light from much more distant galaxies, whose light has been traveling for 12 billion years.

# HOW DO WE KNOW THAT THE UNIVERSE IS EXPANDING?

One of the most important discoveries that has come from studying the early universe and how it has evolved is the principle that the universe is expanding with time—and the longer we can wait around, the faster that expansion is proceeding.

It was not actually that long ago that scientists thought the cosmic family tree ended with our own galaxy—it was believed that all the objects in the night sky belonged to the Milky Way. In the early years of the twentieth century, this view began to change rapidly, and the family tree began to be populated with other galaxies. The work that was done to establish that the Milky Way did, in fact, have siblings, marked the beginnings of a new branch of astronomy—the study of other galaxies beyond our own.

Scientists began to take measurements of what we now know were other galaxies, trying to determine how far exactly they were away from us. However, a curious thing appeared in every measurement of these spiral and elliptical "nebulae," as they were called then. Each one appeared to be moving away from us.

## SHIFTING COLORS

This particular measurement is a relatively straightforward one to undertake. The new stars that form within galaxies are usually surrounded by gas that didn't quite make it into the formation of the star. The newfound star can then give energy to the gas. This energized gas glows a very specific color, depending on what it's made of. Hydrogen, for instance, has a very bright, unique pink-red glow, and oxygen glows green in these environments. The colors are due to the exact wavelength of light that the atoms produce when an electron loses energy, and we know these wavelengths quite precisely from experiments here on Earth.

If the light reaching us from another galaxy is not exactly the color we would expect if we were measuring it on Earth, it means that there's been some movement between the object that

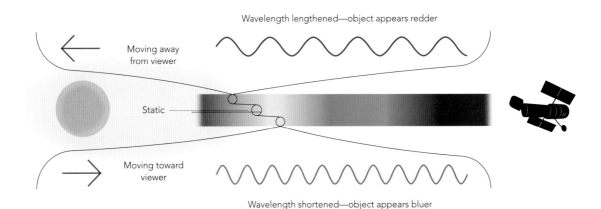

Wavelength lengthened—object appears redder

Moving away from viewer

Static

Moving toward viewer

Wavelength shortened—object appears bluer

produced the light, and us receiving it. If it's moving toward us, the expected colors will arrive bluer than they would have if there wasn't any motion. The peaks and troughs of the light wave are pushed closer together, making them a higher frequency, which registers them as a different color—in this case, blue. If the light wave is stretched out instead, the light will be shifted toward the red. Cunningly, we have termed these changes redshift and blueshift. It's exactly the same phenomenon as a doppler shift in sound waves, changing the tone of a siren as it approaches and then passes you.

So when astronomers looked out at the night sky, and found that everything was redshifted and not blueshifted, that implied that everything was moving away from us. And if everything was moving away from us, the universe couldn't be in balance. If the universe was in perfect balance, with no expansion or contraction, you wouldn't expect to see everything else fleeing from you as though you were at the top of a hill, watching everything roll away.

So either the universe wasn't in balance (in this case it would have to be expanding), or Earth, our vantage point, was in a special place, like the top of our metaphorical hill. Historically, scientists haven't liked arguments that mean we have to exist in a statistically improbable location in the universe, and nowadays we have another argument at our sides that works against this. No matter where we look in the sky, everything looks pretty much the same.

Sure, the small things change—there might be five galaxies in a group here, 13 in a group there, but on average, galaxies are distributed fairly evenly across the sky. And, looking at the distribution of galaxies, it shouldn't matter where we're standing; if we were in a totally different galaxy, in a vastly different part of the universe, we should still see a universe that looks pretty much (at least on a statistical level) the same as ours. The technical terms for this are "homogeneous" (the same from any point you could be standing) and "isotropic" (the same in every direction you look).

◄ **The effect of motion on light**
Objects moving toward us have their wavelength compressed, shifting the color of light toward the blue. Objects moving away from us, on the other hand, have their wavelength stretched out, shifting the color toward the red. In the diagram, the central circle (marked static) marks the colour we would expect from some feature of light (perhaps the presence of sodium) if it were not moving, relative to Earth's location. If, however, the sodium-bearing star is moving toward or away from Earth, we'd expect to see the wavelength, and thus the color, to shift. The lower circle would indicate a blueshifted color (toward blue) and the upper (toward red) would indicate a redshift.

## COLOR COMBINATIONS
The set of colors that each element produces when it's energized by a star is its spectrum, and the spectrum is one of the major tools we use in order to figure out what things are made of. Each element has a unique combination, and it can both absorb away or glow with these unique colors. Because the set of colors is dictated by the structure of the atom, these colors shouldn't change. If their appearance changes, that's due to an outside influence, not to a change in the element itself.

# HOW IS THE UNIVERSE EXPANDING?

The combination of the galaxies in our universe all appearing to recede from us, and the knowledge that our position in the universe isn't remarkable, leads us to another question. If we're nowhere special, how is it that everything appears to be receding away from us? How is the universe expanding? Is there a "center" somewhere that is feeding new space into our universe, or is there no center, and every galaxy is simply embedded in an evenly expanding universe?

Fortunately, these two scenarios would reflect their differences in the way we see galaxies moving away from us, and in the distribution of galaxies in the sky. If the universe is expanding evenly, from everywhere, gradually extending the space between galaxies, then we would actually expect every galaxy to appear to be moving away from us. As a result, this is our current theory of how the universe's expansion is unfolding. If all things are separating, but we are stuck in a fixed location and not moving, then relative to us, all things are moving farther away. Notably, this would appear to be true no matter which galaxy you happened to find yourself in—you would always observe everything to be receding from you. If, on the other hand, space spooled out from some kind of vortex over to the left, we'd notice a difference in the way that galaxies were spread in the universe, and a difference in redshifts as we looked around at different parts of the sky.

Say we're looking directly at our spooling space vortex, and it's pushing new space out into existence. That wouldn't change the distance between our galaxy and the galaxy behind us, away from the vortex. So the galaxy behind us would appear to be stationary relative to us. In fact, most galaxies on "our side" of the vortex would appear to be moving either slowly or not at all, relative to us. Things on either side of the vortex wouldn't appear to be drifting forward or backward, but they'd have some pretty solid apparent sideways motion—at high speeds, this

might drive a measurable parallax (see the box on the next page), where you could see its motion against the background galaxies. Meanwhile, galaxies on the other side of the vortex would appear to recede from us very rapidly!

We'd also notice changes in the distribution of galaxies—and that's because in this scenario, the vortex is forming empty space, so it's creating a perfect bubble: an absolutely empty sphere of space, growing rapidly with time. And if we were to take large-sky surveys of the galaxy population, such a bubble would surely stand out as unusual. We've actually looked for signs that we might be inside such a cosmic bubble (though, clearly, it's not as empty as the vortex-bubble would be), and so far we've found not even a hint of evidence that we're in one.

Instead, data like galaxy counts, temperature evolution over time, and density measurements out to very large distances point toward the expanding, isotropic, homogeneous universe picture, and not to Earth being at the privileged center of a vortex (or explosion). If there were a cosmic vortex somewhere, spooling out space-time from a specific point, ultimately we would expect the way we observe the universe to be very different. Instead, because space is being created between all the objects that already exist, we observe all galaxies to be moving away from us. We also observe that the universe is distributed evenly across the sky, so no matter where we stand, we should see something similar.

## ▼ Measuring distances in space

There are many ways to measure a distance in space; which you use depends on how far you need to see. Very near Earth, parallax measurements are possible, which is useful for mapping the nearby parts of the Milky Way. Farther out, variable stars can be used as standard measurements out to the galaxies closest to the Milky Way. Beyond that, the constancy of the brightness of a supernova detonating a white dwarf indicates a distance. Each method overlaps with the closer ones, allowing us to set up a conversion from one to the other.

## PARALLAX

Parallax is an effect by which nearby objects will appear to move against distant objects in the background, which appear to remain stationary. Any two lines of sight can produce this effect. If you hold a finger out at arm's length with one eye closed, and then close that eye and open the other, you will see your finger appear to jump sideways against the background beyond it. In astronomy we typically use the position of Earth six months apart (so that Earth has moved by twice the distance between Earth and the Sun) to spot the motion of nearby stars against the more distant stars that remain unmoving.

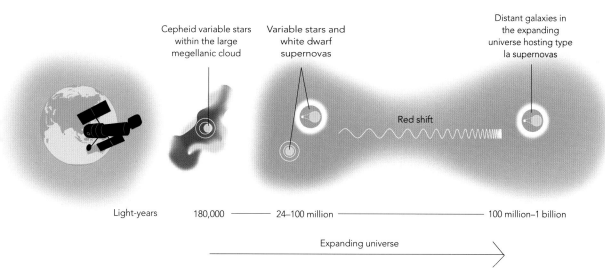

Cepheid variable stars within the large megellanic cloud

Variable stars and white dwarf supernovas

Distant galaxies in the expanding universe hosting type la supernovas

Red shift

Light-years     180,000 ——— 24–100 million ——————— 100 million–1 billion

Expanding universe

# HOW IS IT POSSIBLE TO LOOK BACKWARD IN TIME?

If the universe is expanding, rewinding time should produce a smaller, more compact universe. This is indeed the case, and if we combine this feature with the limitations of the speed of light, it means that the most distant light we're able to observe is also coming from a period of time when the universe was significantly more dense than it is now.

If we think about the observable universe as a series of shells surrounding Earth, where each shell gives us a different window in time on our changing universe, then another feature of this time-traveling view is that the volume of space we can observe with each shell increases the farther back in time we go.

Light's shift toward the red—used to measure whether objects are moving away from us, and if so, at what speed—is useful here. Intrinsically, the redshift of an object tells you how far the light has been shifted from its rest position. But since the most distant galaxies are also the ones that have the largest shifts to their light, redshift can also be used as a metric for the age of the universe. Since we've figured out how fast the universe expanded, and when that expansion happened, we can convert the age of the universe (from redshift) into a rough size of the universe when the light left a distant object. At a redshift of two (see the box on the next page), the universe is a little less than 3.5 billion years old (light from objects at this distance has been traveling for a little over 10.5 billion years) and one third its current size. A galaxy at a redshift of nine is living in a universe of one tenth its current size. Add one to the redshift, and then make that into your fraction. (This math is a bit of a rough estimate, but it's a good way to get a general perspective.)

## DENSE AND DISTANT

Since we are looking back to a universe that is physically smaller, this means that the distances between galaxies are all smaller, and the entire universe at that earlier time must be more dense than it is now. But the critical thing to consider here is the volume of space we're able to observe. We can only see things that are very near to us within a very small volume; at greater distances we see a much larger volume of space. This means that as we look back at more distant objects, we are seeing a larger fraction of the universe, the farther back in time we look. Since we don't know the total volume of the universe, it's impossible to say how much that fraction changes, but it's certainly a bigger number than for the nearby universe!

Our Milky Way's physically closest galactic family members fill only a very small fraction of the current universe. If we try to push to galaxies beyond this immediate neighborhood, we recede backward in time, but we expect that the volume of space that operates close to our "now" should be repeated many times over to fill the universe that we cannot observe but know must be there. Another galaxy's family neighborhood may be filled with more galaxies, or fewer companions, or no companions at all.

This shifting area that we can survey in the universe becomes convenient for the sorts of galactic family censuses that scientists might like to undertake, to try to count how frequently certain types of galaxies or configurations of galaxies appear. These studies of the galaxy population are much easier with large numbers of galaxies, and it's easiest to catch large numbers of galaxies when you have a large volume to look into. The very local universe doesn't afford us the luxuries of large volumes, but at the farthest distances, we can start to say how common groups of galaxies should be.

It's much harder to say whether or not it's uncommon for a galaxy as large as the Milky Way to have as large a companion as Andromeda, given the age of the universe. Our observable universe only has a few galaxies that are close enough to be in the very recent past as we look at them. The shell of space that surrounds us, before the time lag from light's travel time begins to be large, is reasonably small. We can't go hunting for other Andromedas around other Milky Ways in the universe "now," because we begin to go too far back in time the more distantly we search.

## REDSHIFT

Redshift is a ratio of the shift of the light as observed, relative to the expected wavelength of the light if the object weren't moving. For instance, if the wavelength of light in a lab was 656 nanometers, but it had been shifted to 1,968 nanometers because of the expansion of the universe, the difference from the expected is 1,272 nanometers. Dividing out by 656 gives us a redshift of 2.0.

▼ **Time lag**
The time lag of the very nearby universe (not drawn to scale). Our position on Earth lags 8 minutes behind whatever is happening on the Sun; just over 4 hours behind events on Neptune; and 4 years behind the nearest star, as light must take time to traverse the distance between them and us. The farther away from Earth we go, the more out of date our information becomes.

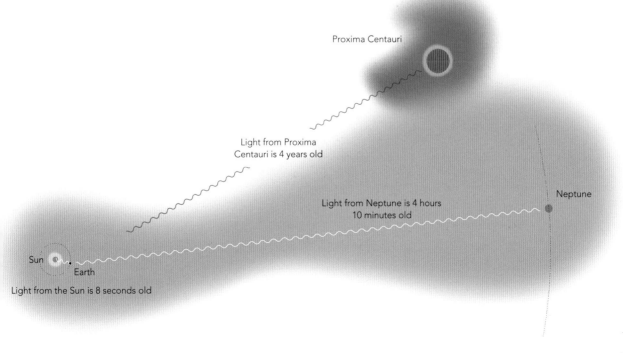

Proxima Centauri

Light from Proxima Centauri is 4 years old

Light from Neptune is 4 hours 10 minutes old

Neptune

Sun
Earth
Light from the Sun is 8 seconds old

# ARE ALL GALAXIES SEPARATING?

The galaxies within the universe are not evenly spaced out. Ours is a universe that looks rather spidery, with filaments of the cosmic web stretching across the sky. There are grand areas of nothing, and clusters of thousands of galaxies, connected to other clusters with faint tendrils, each made up of glowing galaxies.

When we say the universe is expanding, we do in fact mean that the space between objects is increasing, but the weblike distribution of galaxies means that this is not quite as simple as extending the distance between all objects in the universe.

There are two reasons why this is so. The first is that the force driving the universe's expansion appears to be very small. There's an awful lot of space in the universe, so as a bulk property of the universe, we measure a significant change, but in smaller regions of space the effect isn't as strong. The second reason is that in those smaller spaces, the force of gravity is much stronger than the force pushing objects apart. Groups of galaxies, galaxies themselves, and certainly solar systems and planets are not pulled to larger sizes by the expansion of the universe at large because the force of gravity holds them much more tightly together.

Gravity is a bit of a juggernaut in the astrophysical world, and if two objects are gravitationally bound to each other (meaning their relative speeds are too slow to let each escape the pull of the gravity of the other), cosmic expansion isn't going to be able to do much about it. The expansion of the universe would have to be frighteningly enormous (much larger than we observe it to be) to pull apart two galaxies that are bound to each other by gravity.

In the last chapter we talked about the future collision of the Milky Way and Andromeda. We can say with such certainty that they will collide because they are gravitationally tied to each other—the expansion of the universe will not draw

## The force driving the universe's expansion appears to be very, very small.

the two of them apart. Any bound objects in the universe won't be affected by this expansion, so not only are the Milky Way and Andromeda immune to their universal parent's continual expansion, but the entire cluster of galaxies to which our galaxy belongs will remain in place. However, galaxies in a neighborhood of the universe that are not gravitationally tied to our galaxy or its cluster of companions will find themselves drifting ever farther from us over time.

▶ A visualization of the Millennium Simulation, which simulates the cosmic web by tracing how material collects over time. This simulation used a month of supercomputer time to complete the image, and illustrates the filamentary nature of the universe—galaxies tend to collect where material is the most concentrated. Clusters of galaxies might be expected to form at the bright yellow dots where filaments cross.

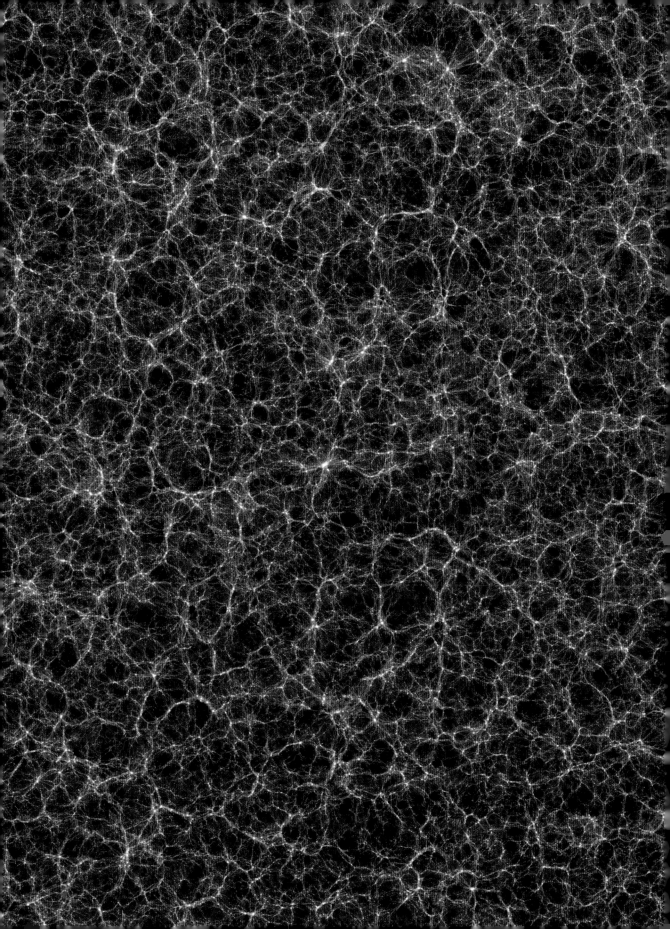

# WHAT'S PLANET EARTH'S ADDRESS IN THE UNIVERSE?

To make a usable map of our cosmic lineage is quite a tricky thing. From our vantage point on Earth, we have mapped out a list of relationships between our planet and the others within our solar system, between the Sun and the other stars, and between our Milky Way and the other galaxies. With every tier of our family tree we have tied our understanding to our own home, whether that be planet, star, or galaxy.

To go forward from a description of family ties to a detailed mapping of where we actually stand, however, requires not just information on respective positions, but measurements of the distances to each of our family members as well.

We know our cosmic lineage quite well up to a certain point. You, reading this book, have a street address in a country on planet Earth. Earth is part of our solar system of planets, which orbits the Sun. The Sun is one of many billions of stars that orbit the center of the Milky Way galaxy. The Milky Way, in turn, orbits the center of mass of the Local Group, somewhere in between the Milky Way and Andromeda. The Local Group sits within the Local Supercluster, a loose association of other clusters and groups of galaxies, spanning hundreds of millions of light-years.

But this kind of lineage is only useful if you know where to find the Supercluster. If you know where any individual part of our lineage is located, you might be able to trace your way back to find our planet from elsewhere in the universe, but without that anchor to a different set of knowledge, it's not particularly helpful or informative.

## RETURN TO SENDER
Scientists have grappled with this problem on a much smaller scale before. Both the *Pioneer* and *Voyager* spacecrafts tried to relay our position in the unlikely event that they were ever encountered and picked up by another intelligent race of beings. The *Pioneer* plaque showed the location of our planet within our solar system, but it (and the *Voyager* golden record) also provided a map that would be useful to someone else from within our galaxy. The set of lines radiating out from a single point on the left-hand side of the *Pioneer* plaque and the *Voyager* record is this map. This gives the positions and frequencies of a set of pulsars—incredibly rapidly spinning objects that beam light out into the cosmos like a lighthouse. Each pulsar has its own unique frequency: the number of times per second it flashes in our direction. Pointing out the distances and frequencies of 14 pulsars, as we did, should allow someone else, observing the same set of pulsars, to triangulate our position.

However, this can only be scaled up so far. If you're trying to tell someone outside our galaxy where we are, you swiftly become tangled in a different problem, which is that the farther from our planet you would like to map out for our faraway friends, the more you travel back in time. At some "distance" from our planet, our exploration becomes one much more of time than of space. Even the Virgo supercluster (a tiny corner of the universe) is some 110 million light-years across. Between light leaving that side of the cluster and our receiving it, things have changed over there in the intervening 100 million years. On an astronomical timescale, 100 million years is not a huge amount of time, but it's much longer than humans have inhabited Earth.

This time traveling becomes an increasingly disruptive problem the farther back in time we go. At some point, we are looking at structures and objects that no longer exist. Many things can happen in a couple of billion years, and they usually do, so using those objects on our map isn't very useful for an intergalactic or universe-wide traveler.

Unfortunately, there's no solution to mapping out our lineage in a foolproof manner so that anyone, anywhere, would be able to point to the planet where the humans reside. One of the fundamental tenets of cosmology tells us that there are no "special" perspectives on the universe. This means that there's no overall frame of reference that every possible observer could agree on, no universe-wide "north" by which to orient our maps. The best

thing we may be able to do right now is to create a really good map of our little local set of galaxies and clusters, and tell any visitors, if they happen to see something that looks like it in their travels, that we humans are the children of this particular part of the universe.

▼ The plaque attached to the *Pioneer* spacecraft. Along the bottom is a map of our solar system, showing the path the spacecraft took to leave the solar system. On the right, humans are illustrated to scale with the spacecraft behind them. The pulsar map to indicate the location of our solar system is on the left.

# If the universe were tiger-shaped, how would we know?

Children sometimes invent incredible cosmologies, and a tiger-shaped universe is one of them: "If the universe were in a giant tiger, we could—with a spaceship and enough gas—get out through its mouth." Tigers don't match our current understanding of the universe, but why not? Our current theories indicate that the universe is geometrically flat, spatially infinite, but finite in time. Where did these theories come from, and which observations rule out a cosmic tiger?

Given that there's nothing ruling out a grand cosmic tiger birth, we'll pass over the "finite in time" principle. However, a tiger-shaped universe would not align with a geometrically flat, spatially infinite universe. You might be wondering, then, why we are so certain that these two should hold. Let's look at both these concepts in more detail.

### 1. Why assume flatness?
We humans usually operate on assumptions of geometric flatness. A piece of paper is geometrically flat; draw a triangle on it, and add the angles in the corner, and you'll get 180 degrees. You can curl the paper into a tube, but there's only so many shapes you can make without cutting or folding the paper. If your paper is infinitely large, you can start drawing two parallel lines on it, and they will extend out into infinity, never crossing or looping back on themselves.

By comparison, Earth has a different geometry. It's a sphere. If you trace a small triangle, you'll get an angle that is pretty close to 180, but the larger you draw your triangle, the larger your angles get. At an extreme, you could have one edge of your triangle along the equator, and then go due north from two points, which is a 90-degree turn, and then have those two points meet at 90 degree angles at the North Pole. This would trace out a quarter of the northern hemisphere. Lines drawn on a circle typically cross each other twice, and they loop back on each other. This geometry is fundamentally distinct from a flat one; try to fold a piece of paper into a perfect sphere and you'll find yourself frustrated quickly. Similarly, if you tried to deflate a sphere (a globe, or a ball, for example) and fold it flat, you'd find yourself with a wrinkled surface that just won't cooperate.

There's one other potential geometry, and it's called a hyperbolic surface. Roughly speaking, it has the shape of the kind of curved crisp that is usually sold in a tube. In one direction, the surface curves upward, and in the other direction, downward. This is dubbed a saddle shape. In this geometry, you also can't get a piece of paper to cooperate without cutting it into a new shape, nor would you be able to smash your crisp flat without breaking it. A large enough triangle in this universe would have smaller than expected angles in its corners.

A tiger-shaped universe would be geometrically complex, and it most certainly would not be flat. Imagine taking all the stuffing out of a plush tiger; you might be able to get some parts of it to lay flat, but certainly not all of it. Tiger armpits would act like a hyperbolic surface, and the tail would be a cylindrical universe.

Observations so far have told us our universe is flat, or in any case so close to flat that we can't measure a difference from flat. This certainly excludes strong curvatures, as would be required to carve out a cosmic tiger. And it is flat in every direction we look; which also excludes strong changes in curvature (say between a tail shape and a leg shape).

## 2. Why spatially infinite?

Critically, the amount of tiger in different directions would not be the same. Say we sat in the exact middle of the tiger belly, looking outward in the tiger universe. We'd expect to see more galaxies in the directions of paws and tail, since there's more space in that direction, and fewer galaxies if we looked up toward the back or sides of the tiger.

▼ The three potential geometries of any universe are drawn below. Flat geometry is the most familiar, but there are other options. Hyperbolic and spherical shapes are also possible. Two beams of light, shining outward, would behave differently in each universe.

Our own observations of the universe are that there is the same number of galaxies in every single direction we look. This tells us we are either in the exact center of a spherical universe, or else the universe is spatially infinite. The first is so profoundly unlikely that we've ruled it out. Why should Earth be in a special location in the universe? If Earth is not in a special location, then any observer should see roughly the same thing. And if *any* observer should see roughly the same thing, then the universe has to continue filling every direction with an infinite number of galaxies.

### Verdict

A tiger-shaped universe is, alas, incompatible with our observations of the universe. Our observations tell us we must be in a geometrically flat, spatially infinite universe, where triangles add up to 180 degrees, light can travel forever in one direction without looping and we sit in no special place within it.

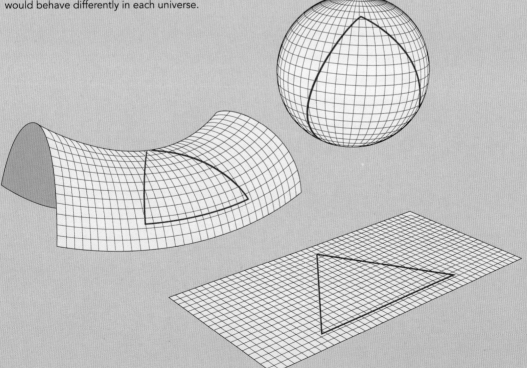

# INDEX

# PICTURE CREDITS

7: (from top): © milanares/Adobe Stock; © NASA/WMAP Science Team; © NASA, ESA, Hubble Legacy Archive; © NASA & ESA; © Pablo Carlos Budassi/CC A-SA 4.0; © The Hubble Heritage Team (AURA/STScI/NASA); © lukszczepanski/Adobe Stock; © max dallocco/Adobe Stock; bottom & 131: © Keesscherer/CC A-SA 4.0

8: © Dr_Flash/Shutterstock

12: © NASA

13: © NASA/ESA/M. Robberto

15: © NASA, ESA, and M. Livio, The Hubble Heritage Team and the Hubble 20th Anniversary Team (STScI)

18: NASA/M. Ahmetvaleev

21: © Mike-Hubert.com/Shutterstock

22: (ISS) © Nerthuz/Shutterstock; (Earth) © sdecoret/Shutterstock

25: top © NASA; bottom © ESA, CC A-SA 3.0 Contains modified Copernicus Sentinel data (2019)

26: © NASA/Earth Observatory

27: © NASA/GSFC/METI/ERSDAC/JAROS, and U.S./Japan ASTER Science Team

29: © ESO/Yuri Beletsky/CC A-SA 4.0

30: © NASA

31: © ESO

32 & 43 bottom left: © NASA/Bill Ingalls

35: © NASA/John Kaufmann

37: © NASA/NOAA

38: both © Vintagepix/Shutterstock

39: © NASA/NOAA

40: © NASA/JSC

43: top © NASA/Paul E. Alers; bottom right © NASA/MSFC/Joe Matus

45: © NASA/DOE/Fermi LAT Collaboration

46: Diagram © NASA's Imagine the Universe; telescope image © the HESS Collaboration

47: © NASA / Neil A. Armstrong

54: © NASA/JPL-Caltech

55: © NASA/JPL-Caltech

57: Venus © magann/Adobe Stock; Mercury & Jupiter © mode_list/Adobe Stock; Earth max dallocco/Adobe Stock; Mars © dottedyeti/Adobe Stock; Saturn © Tristan3D/Adobe Stock; Neptune & Uranus © NASA/JPL

59: © NASA/JPL-Caltech/ASU

60: top © Bjoern Wylezich/Shutterstock; middle © H. Raab/CC A-SA 3.0

61: © NASA/JPL-Caltech/Cornell University

62: © Marie-Lan Taÿ Pamart/CC A-SA 4.0

63: © ESA/Hubble & NASA, Z. Levay

64: left © PHOTOOBJECT/Shutterstock; right © Dr M.Rohde,GBF/Science Photo Library

65: © Harel Boren, CC A-SA 4.0.

66: © NASA/JPL-Caltech

67: © NASA/JPL/USGS

68: © NASA/JPL

71: © NASA/JPL-Caltech/MSSS

73: © Enhanced image by Kevin M. Gill (CC-BY) based on images provided courtesy of NASA/JPL-Caltech/SwRI/MSSS

74: © NASA/JPL-Caltech/SwRI/MSSS/Gerald Eichstadt/Sean Doran © CC NC-SA

75: top © Enhanced image by Kevin M. Gill

(CC-BY) based on images provided courtesy of NASA/JPL-Caltech/SwRI/MSSS; bottom © NASA/JPL

79: © NASA/Johns Hopkins University Applied Physics Laboratory/Southwest Research Institute

81: top © NASA/JHU APL/SwRI; bottom © NASA/Johns Hopkins University Applied Physics Laboratory/Southwest Research Institute

82: © NASA/Ames/JPL-Caltech

85: top © NASA/JPL-Caltech; bottom: © NASA/JPL

88: © Jim Peaco, National Park Service

91: © NASA/JPL-Caltech

93: © 3d_man/Shutterstock

95: © NASA/Ames/SETI Institute/JPL-Caltech

96: © NASA/Goddard/SDO

105: © NSO/NSF/AURA (CC A-SA 4.0)

109: © NASA

111: © NASA/GSFC/SOHO/ESA

113: © ESA/NASA/Soho

114: © NASA

117: © NASA, ESA, and the Hubble Heritage (STScI/AURA)-ESA/Hubble Collaboration

118: © Rogelio Bernal Andreo, CC A-SA 3.0

119: © NASA/JPL-Caltech

123: © NASA/ESA Hubble Space Telescope

125: © NASA, ESA and H.E. Bond (STScI)

126: © NASA/JPL-Caltech

127: top © Dr. Mark A. Garlick

131: top © Keesscherer, CC A-SA 4.0

133: © NASA/JPL-Caltech/DSS

135: © ESA/Hubble & NASA

137: © NASA, ESA and the Hubble Heritage Team (STScI/AURA). Acknowledgment: A. Zezas and J. Huchra (Harvard-Smithsonian Center for Astrophysics)

140: © X-ray: NASA/CXC/Penn State/S.Park et al.; Optical: Pal.Obs. DSS

147: ESO/Digitized Sky Survey 2. Acknowledgment: Davide De Martin.

149: © NASA/ESA, The Hubble Key Project Team and The High-Z Supernova Search Team

151: top © NASA, ESA and the Hubble Heritage Team (STScI/AURA); bottom: © ESO/L. Calçada

153: © NASA/CXC/Rutgers/J.Hughes et al.

155: © NASA/JPL-Caltech

156: © NASA/Casey Reed

157: X-ray: NASA/CXC/RIKEN/D.Takei et al; Optical: NASA/STScI; Radio: NRAO/VLA

159: © ESO/L. Calçada/M.Kornmesser

161: © NASA, ESA, and D. Coe, J. Anderson, and R. van der Marel (STScI)

164: © Mysid/CC A-SA 3.0

167: © ESO/L. Calçada/M.Kornmesser

169: © NASA, ESA, the Hubble Heritage (STScI/AURA)-ESA/Hubble Collaboration, and A. Evans (University of Virginia, Charlottesville/NRAO/Stony Brook University)

170: © ESA/AOES

176: © NASA/ESA and The Hubble Heritage Team (STScI/AURA)

179: © Steve Jurvetson/CC A-SA 2.0

181: © NASA, ESA, and The Hubble Heritage Team (STScI/AURA); Acknowledgment: P. Knezek (WIYN)

182: © NASA/ ESA/ STScI (S. Beckwith)/HUDF Team

185: both © ESO/VVV Survey/D. Minniti/Serge Brunier; Acknowledgment: Ignacio Toledo, Martin Kornmesser

186: © NASA, ESA, S. Baum and C. O'Dea (RIT), R. Perley and W. Cotton (NRAO/AUI/NSF), and the Hubble Heritage Team (STScI/AURA)

187: © ESO/L. Calçada/spaceengine.org

189: © NASA, ESA, SAO, CXC, JPL-Caltech, and STScI. Acknowledgment: G. Fabbiano and Z. Wang (Harvard-Smithsonian CfA, USA), and B. Whitmore (STScI)

193: © ESO/M. Kornmesser

195: © Adam Block/Mount Lemmon SkyCenter/University of Arizona/CC A-SA 4.0

196: © ESA/Hubble & NASA, S. Jha; Acknowledgment: L. Shatz

200: © NASA/WMAP Science Team

204–206: © ESO/S. Brunier

207: © NASA/ESA

215: 215: © Volker Springel et al. (Virgo Consortium)/Max Planck Institute for Astrophysics/Simulation code: Gadget-2

217: © NASA Ames Resarch Center (NASA-ARC)

219: © NASA / WMAP Science Team

Images not listed on this page are in the public domain. Every effort has been made to credit the copyright holders of the images used in this book. We apologize for any unintentional omissions or errors and will insert the appropriate acknowledgment to any organizations or individuals in subsequent editions of the work.

With the exception of those listed above, all illustrations are by Rob Brandt.

## ACKNOWLEDGMENTS

Thanks to Mum, Dad, Matthew, and to all my friends, who consistently got excited about what I was writing about.

Thank you to my editors, Duncan Heath and Richard Webb, who found all the rough edges of the original manuscript and helped me smooth them out.

And to all the readers of "Astroquizzical" over the years, thank you for giving me your curiosity to satiate.